MICROSCOPY HANDBOOKS 32

Microscopy of Textile Fibres

Royal Microscopical Society MICROSCOPY HANDBOOKS

Microscopy of Textile Fibres

P.H. Greaves

Microtex, 7 Newall Carr Road,
Otley, West Yorkshire LS21 2AU, UK

B.P. Saville

Textile Department, Huddersfield University,
Queensgate, Huddersfield HD1 3DH, UK

Taylor & Francis
Taylor & Francis Group

LONDON AND NEW YORK

Published by Taylor & Francis
2 Park Square, Milton Park, Abingdon, Oxon, OX14 4RN
270 Madison Ave, New York NY 10016

Transferred to Digital Printing 2009

Typeset by AMA Graphics Ltd, Preston, UK.

Publisher's Note
The publisher has gone to great lengths to ensure the quality of this reprint but points out that some imperfections in the original may be apparent.

Contents

Abbreviations

BS	British Standard
CLSM	confocal laser scanning microscope
FDA	fibre diameter analyser
FTIR	Fourier-transform infrared
IR	infrared
ISO	International Standards Organization
IWTO	International Wool Textile Organization
OBA	optical brightening agent
OFDA	optical fibre diameter analyser
OPD	optical path difference
RI	refractive index
SEM	scanning electron microscope
TEM	transmission electron microscope
TSRLM	tandem scanning reflected light microscope
UV	ultraviolet

Preface

Textiles, like food, cosmetics, paper and photographic substrates, do not fall easily into either of the traditional divisions of materials science or biological microscopy. Invariably they possess some characteristics of one school and equally strong ones of the other. Thus, calibrated measurements may be made as often as subjective assessments, brightfield illumination used together with polarized light and thin TEM sections compared with UV fluorescence of the same material.

These factors place textiles, along with the products listed above, in the largely grey area of what may be termed 'consumer science'. The microscopy of this category of materials generally receives far less attention than do the 'classic' fields of metallurgy and biology. Meetings, symposia and courses rarely present significant emphasis on fibre, food or photographic microscopies.

It may surprise the reader unfamiliar with textiles to learn that most large fibre and fabric manufacturers will have at least one form of microscope on site which is regularly used. The instruments can range from the fully equipped light and electron microscopy laboratory of a major fibre producer, to a simple stereo or monocular compound microscope kept by a knitter or weaver.

This book is intended to assist the microscopist who has little or no experience of textile materials, but who may be called upon to examine them.

P.H. Greaves and B.P. Saville

Acknowledgement

The authors would like to thank Dr M.G. Dobb of the Textile Physics Laboratory, University of Leeds for the use of the transmission electron micrographs in Chapter 8.

Safety

Attention to safety aspects is an integral part of all laboratory procedures, and both the Health and Safety at Work Act and the COSHH regulations impose legal requirements on those persons planning or carrying out such procedures.

In this and other Handbooks every effort has been made to ensure that the recipes, formulae and practical procedures are accurate and safe. However, it remains the responsibility of the reader to ensure that the procedures which are followed are carried out in a safe manner and that all necessary COSHH requirements have been looked up and implemented. Any specific safety instructions relating to items of laboratory equipment must also be followed.

1 Introduction

Few who are not involved in the industry would think far beyond clothing when considering the primary uses of textile products. Clothes provide our daily contact with textiles, and most of us will have at least some idea of the fibre composition of our favourite sweater, jacket or coat. This may be limited to the casual association of fibre type with the feel or 'handle' of a garment and whether it is considered 'silky', 'harsh' or 'crisp' by the wearer. The washing procedures and care of garments are further factors which may familiarize the consumer with different fibre types. Clothing, however, is only one use of textiles, and fibres are employed in far more applications than mere apparel.

In the home, floorcoverings, curtains, upholstery, bedding and towelling are some examples, while in transport vehicles the seats, belts, carpets, insulation and tyres depend on textile fibres. Outdoors, geotextiles protect land from erosion and permit growth of plant and tree cover in exposed or barren areas. In industry, textiles are used in conveyor belts, filter cloths, and bagging and sound-absorbing materials. Even clothing itself is more diverse than might first be imagined, and may include underwear, hosiery, shirtings, knitwear, suitings, outerwear, uniforms, footwear and protective garments. Medical and personal hygiene textiles are further examples.

Each of these textile products will be made of specific fibre types according to the properties required in use, and the way in which the fibres are assembled into fabric will further influence the product's performance.

1.1 Fibres

The accepted definition of a textile fibre is 'a unit of matter characterized by its fineness, flexibility and having a high ratio of length to thickness'. Fibres are the basic units from which all textile materials are made.

There are many different types of textile fibre and, for convenience, they are usually classified as either natural or man made in origin. These classes are further subdivided into animal or vegetable for natural fibres,

Table 1.1:   Examples of fibre types and their composition

Generic name	Type	Principal component
Acrylic	Man made, synthetic	Polyacrylonitrile
Angora	Natural, animal	α Keratin (protein)
Aramid	Man made, synthetic	Aromatic polyamide or polyimide
Alpaca	Natural, animal	α Keratin (protein)
Cashmere	Natural, animal	α Keratin (protein)
Chlorofibre	Man made, synthetic	Vinyl or vinylidene chloride
Cotton	Natural, vegetable	Cellulose
Diacetate (acetate)	Man made, regenerated	Secondary cellulose acetate
Elastomer	Man made, synthetic	Segmented polyurethane or polyisoprene
Flax	Natural, vegetable	Cellulose
Fluorofibre	Man made, synthetic	Polytetrafluoroethylene
Glass	Man made, inorganic	Principally silica
Jute	Natural, vegetable	Cellulose
Kapok	Natural, vegetable	Cellulose
Modacrylic	Man made, synthetic	Copolymer of acrylonitrile and vinyl or vinylidene chloride
Modal	Man made, regenerated	Cellulose
Mohair	Natural, animal	α Keratin
Nylon (polyamide)	Man made, synthetic	Polyhexamethylene adipamide or polycaproamide
Polyester	Man made, synthetic	Polyethyleneglycolterephthalate
Polyethylene	Man made, synthetic	Polyethylene
Polypropylene	Man made, synthetic	Polypropylene
Ramie	Natural, vegetable	Cellulose
Rabbit	Natural, animal	α Keratin (protein)
Silk	Natural, animal	β Keratin (protein)
Triacetate	Man made, regenerated	Cellulose triacetate
Viscose	Man made, regenerated	Cellulose
Vinylal	Man made, synthetic	Acetalized polyvinyl alcohol
Wool	Natural, animal	α Keratin (protein)
Yak	Natural, animal	α Keratin (protein)

and regenerated or synthetic for man-made ones. Examples of the different types in each group are explained below, and *Table 1.1* lists some of the more popular kinds of fibre, their type and basic chemical composition.

1.2 Natural fibres

1.2.1 Animal

These fibres are obtained from the coats or fleeces of animals, or in the case of silk from the extruded filaments forming the cocoon of the silkworm. Examples of animal fibres include wool, mohair, cashmere, alpaca, rabbit fur, camel hair, mink, and cultivated and wild silk.

1.2.2 Vegetable

Vegetable fibres grow as seed hairs, bast (stem) fibres, and leaf and fruit fibres. The most common of these is cotton – a seed hair which accounts for almost half of all fibre production world wide.

Other vegetable fibres are kapok, flax, jute, hemp, sisal, ramie and coir.

1.3 Man-made fibres

1.3.1 Regenerated

The term 'regenerated' is applied to fibres which are formed from naturally occurring polymers by modifying and re-forming the original material. Regenerated fibres were once made from proteins (milk, peanuts, soya) but now almost all fibres of this type are based on cellulose. Viscose rayon, the most common, is made from wood pulp. Cuprammonium rayon, cellulose diacetate and cellulose triacetate are also regenerated fibres which are based on cellulose.

1.3.2 Synthetic

Fibres made from polymers produced by chemical synthesis, i.e. where the fibre-forming material has been engineered specifically, are called synthetic or chemical fibres. The raw material for these is oil, and a wide variety of fibre types exist. Nylons, polyesters, acrylics, chlorofibres, polyolefins and aramids are all examples of synthetic fibres.

1.4 Production

The total amount of textile fibres produced in 1989 was 37 939 million kilograms, excluding approximately 12 000 million kilograms of polyolefins (IWS, 1990). This quantity was not divided equally between the different fibre types and, as has long been the case, was dominated by two particular classes of fibre – cotton and man-made synthetics. The approximate share of the different types of fibre is shown in *Figure 1.1*.

In summary, from this chapter it may be seen that textile fibres exist as numerous types, have almost infinite uses and are produced in enormous quantities world wide. They are clearly of major importance to society.

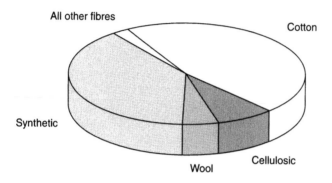

Figure 1.1: World fibre production, 1989.

Reference

IWS. (1990) *Wool Facts*. IWS Development Centre, Ilkley, UK.

2 Fibre Identification

The determination of generic fibre type has always been an exercise of prime importance to those involved with textiles. To the technologist, the selection and purchase of raw material, mechanical and chemical processing, coloration, finishing, and suitability for end use are all governed by fibre composition. The consumer chooses textile goods according to their style, price and fibre content. In the case of rare or prestigious fibres, the fibre content is often more important than either of the other two considerations. This can also be so with items of relatively high cost – e.g. carpets and upholstery. Virtually all textile products are required by law to display a label which accurately declares their fibre contents. The authenticity of these labels is checked by trading standards officers or their equivalents, who will often make retail purchases of goods with the specific aim of having their fibre compositions established by independent testing laboratories. If goods are labelled incorrectly the consequences for those responsible are serious. Customs officers, historians, forensic scientists, launderers and insurance claim investigators are further examples of those for whom fibre identification is an essential task. Although it may be said that because of the wide range of techniques which may be used to identify fibres there is no single 'standard' method, light microscopy is by far the simplest and most often applied.

2.1 Sampling

Unless only a single fibre, or a specimen of such small size that it is self limiting, is the material to be identified, the first step towards accurate fibre identification is to obtain a reasonably representative sample for examination. Textiles are notoriously variable materials, and are often produced as blends of two or more different fibre types. If there is any doubt that the item to be identified is composed of one single type of fibre, then some form of sampling must be carried out prior to examination. Because the homogeneity of any blending will be determined by the stage of manufacture and class of end product, different sampling procedures are used for each state of assembly. The following guidelines summarize the principles of sampling, assuming sufficient material is available.

2.1.1 Loose fibre

The bulk sample is divided into a minimum of 32 zones, each of which is then divided into approximately equal parts. A table of random numbers or a pack of playing cards is used to determine which half of each zone is discarded, e.g. an odd number would cause the left hand half to be discarded, an even number the right hand half. The retained portion of each zone is again divided and half discarded. The sequence is continued until only a small quantity of fibres per zone are retained. When all the zones have been reduced in this way the total retained fibres form the test specimen, which is representative of the bulk sample.

2.1.2 Yarn

If the bulk sample consists of packages of yarn (e.g. cones, bobbins, cheeses etc.) several 1-m lengths are wound off from each and laid together parallel to form a tow. If the yarn is a single hank or skein it is cut into as many 1-m lengths as possible and these are laid parallel, as above. The parallel yarns are now sampled by taking several lengths of approximately 10 mm to produce a composite representative sample.

2.1.3 Fabric

Samples of warp and weft yarn are taken separately from at least three different areas of the fabric, ensuring that all colours and types of yarn present are included. The warp and weft yarns are treated as separate samples, and any selvedges are excluded. The collected yarns should be cut to 10-mm lengths prior to mounting.

2.1.4 Made-up articles

It is first necessary to determine whether all parts of the article are of similar composition. If so, the whole article is treated as the laboratory bulk sample and the procedure for fabric (above) is followed. If parts of the article are of different composition, the parts are separated and each is treated as a laboratory bulk sample of fabric, again as above.

These outlines are intended to convey the general principles of obtaining representative samples from textile materials for fibre identification. For more detailed sampling procedures, the reader is advised to consult British Standard (BS) 4658 (1978) or its international equivalent International Standards Organization (ISO) 5089 (1977).

2.2 Mounting

Having composed a representative sample of the material to be identified, it may now be prepared for microscopical examination. The majority of textile fibres have refractive indices (RIs) of between 1.5 and 1.7, and it is therefore necessary to mount them in a medium which is sufficiently close to these values to permit clear examination, without excessive refraction effects at edges, but not so close as to render uncoloured fibres invisible in normal light. The other properties required of a mountant are that it is stable, non-swelling, colourless and safe. Liquid paraffin of RI 1.47 fulfils these requirements, and has the added advantage of being cheap and readily available. Only cellulose diacetate and triacetate fibres of RI approximately 1.46–1.47 are not clearly visible in liquid paraffin and, if their presence is suspected, a second preparation using water or cedar wood oil as the mountant should be made.

To prepare a slide of fibres for examination, a few drops of liquid paraffin are first placed on to a slide. If loose fibres are to be the test specimen, they are placed in the paraffin and then teased apart with dissecting needles to cover approximately the area of a coverslip. Yarns to be examined should be laid parallel but spaced (about three 10-mm lengths is ideal for one slide). The ends of the yarns are then teased apart so that the individual fibres may be seen. It may be necessary to add more mountant at this stage if it is absorbed by the fibre or yarn assembly. Next, the coverslip is carefully lowered on to the preparation by putting one edge on the slide while supporting the other over the mountant with a dissecting needle. As the needle is lowered, the coverslip contacts one edge of the mountant and forms a meniscus which excludes air bubbles as the slip is slowly lowered to rest on the preparation (*Figure 2.1*).

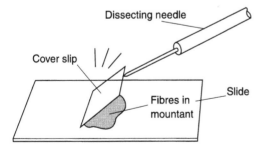

Figure 2.1: Lowering the coverslip to complete the slide.

2.3 Conditions for initial examination

The ideal instrument to use for the first stages of fibre identification is the ordinary transmitted light compound microscope. No great magnification is necessary and a 10× achromatic objective and 10× eyepieces should be quite satisfactory. When examining longitudinal whole mounts of fibres, the overall shape and surface structure is generally of more immediate interest than the fine, internal details. For this reason it is often convenient to increase the depth of field of the system by lowering the substage condenser and stopping down the aperture iris slightly. Simple experimentation will provide the optimum conditions for the particular specimen and Köhler illumination can soon be restored when higher resolving power is needed or thin specimens such as cross-sections are to be studied. Even with the extended depth of field suggested, it will be found that constant adjustment of focusing is necessary to view the many different fibres on a slide, particularly at cross-over points, and when crimped or textured fibres are present.

The likelihood of more than one fibre type being present in an unknown specimen, and the possibility of a component constituting from 0 to 100% of the overall composition of a material means that a cursory examination of the prepared slide will not be sufficient. Rather, the whole slide should be systematically scanned and each different class or characteristic of the fibre be noted.

2.4 Reference samples

Before any serious attempt is made to identify the fibres present in a material of unknown composition it is well worthwhile for the operator to become acquainted with the basic features of authentic specimens. No great effort is required to obtain samples of the most common textile fibres because, as stated earlier, all goods from reputable suppliers will have their fibre contents clearly declared on permanent labels. Even if uncertainty exists, it is still possible to assemble a relatively reliable collection of samples, and to check the composition of these by referring to the known microscopical characteristics of the supposed fibres present. A good example is that of men's shirts, which today are often made of a polyester/cotton blend – and are clearly stated as so. Ladies' fine hosiery is usually composed of 100% nylon, while curtain fibres should never contain nylon because it progressively degrades in sunlight. Tea towels are good sources of flax, the processed term for which is linen, and cotton wool, despite its name, is largely composed of viscose. Bath towels provide a good source of pure cotton, as do many bedsheets and sewing threads. Wool may be found in knitwear (as may acrylic), high class suitings and genuine lambswool paint rollers and polishing cloths. There are many more examples and, if labels are checked and a few fibres plucked from the surface of the item examined for confirmation, it is easy to compile a helpful set of reference slides or samples. Consulting these briefly before examining an unknown can be of tremendous help to the less experienced fibre analyst.

2.5 Key features

Rather than aiming for an instant identification of the precise fibre types present in a specimen, it is recommended that, if possible, the fibres are first classified into their correct groups. *Figure 2.2* in the form of a flow chart shows how this may be achieved, while *Figures 2.3–2.7* illustrate some of the basic features referred to.

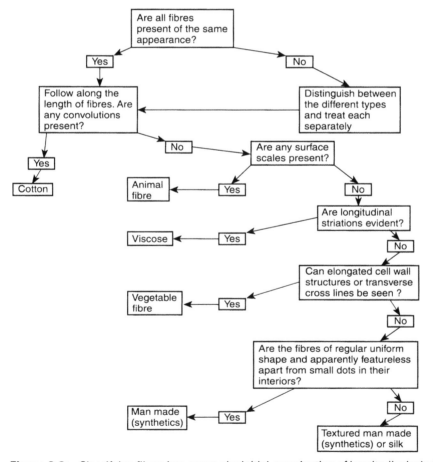

Figure 2.2:   Classifying fibres into groups by initial examination of longitudinal whole mounts.

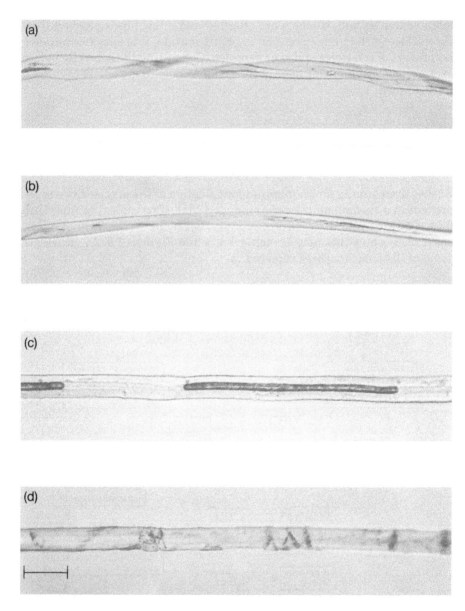

Figure 2.3: Natural fibres: vegetable. Bar = 100 μm. (a) Cotton, unmercerized, (b) cotton, mercerized; (c) jute, ultimate; (d) flax, ultimate.

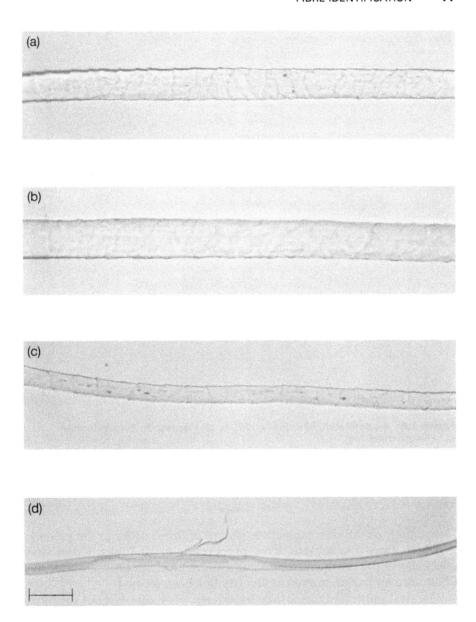

Figure 2.4: Natural fibres: animal. Bar = 100 μm. (a) Wool; (b) mohair; (c) cashmere; (d) silk (*Bombyx mori*).

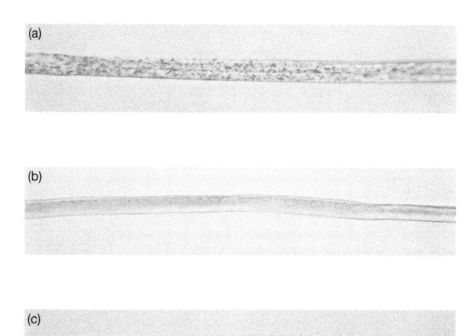

Figure 2.5: Regenerated fibres. Bar = 100 µm. (a) Viscose; (b) modal viscose; (c) cellulose diacetate.

2.6 Solubility tests

Due to their chemical composition, groups of fibres differ in the solvents in which they are soluble. This property may be investigated very effectively under the microscope using a minimal amount of representative fibre, once an initial opinion of identity has been formed.

The most important factor to be considered when carrying out solubility testing is that of safety – both of the operator and the microscope. Although only fractions of millilitres of solvents are used, their ability to dissolve

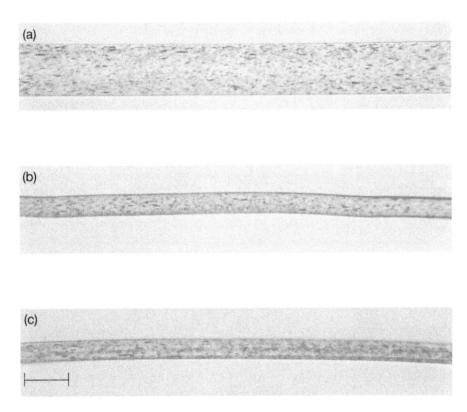

Figure 2.6: Synthetic fibres. Bar = 100 μm. (a) Nylon (6 denier); (b) polyester; (c) acrylic.

fibres means that they are invariably corrosive and often toxic. **When using such solvents, utmost care should be taken to avoid eye contact, skin contact and inhalation of vapours. Suitable protective clothing (gloves and safety spectacles) should be worn and vapours removed from the vicinity of the microscope during slide preparation.** Of course the microscope itself can also be affected by these solvents, and its optical and mechanical parts may easily become damaged if errors are made. Nevertheless, if sufficient care and precautions are taken, solubility tests offer an invaluable means of confirming fibre identities.

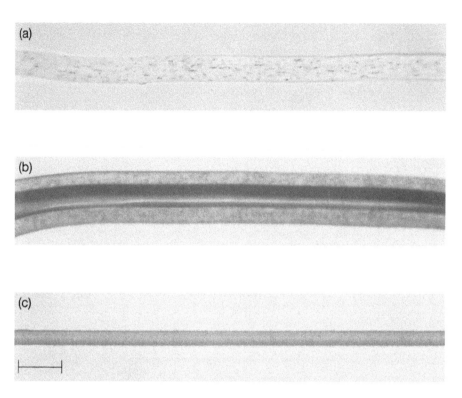

Figure 2.7: Speciality synthetic fibres. Bar = 100 µm. (a) Polypropylene; (b) polyester (hollow); (c) aramid (Kevlar).

The basic procedure for carrying out a solubility test is as follows.

1. Prepare a dry mount of the fibres in question in the same way as for the preliminary examination.
2. Omit the mountant but apply the coverslip.
3. Place the dry slide carefully on the stage of the microscope and locate the fibres to be tested.
4. Using a fine pipette, slowly run under the edge of the coverslip just enough solvent to fully immerse the fibres. Remove the pipette to one side (safely) and quickly observe the behaviour of the fibres in the solvent. Continue observing until the fibres either react with the solvent (i.e. dissolve or swell) or remain unchanged. Ten minutes is a usual observation period, though many reactions are much quicker.

Table 2.1 shows examples of solvents normally used in solubility tests.

Table 2.1: Examples of solvents used in solubility tests for fibre identification

Fibre type	Soluble in:	Insoluble in:
Acrylic	Hot γ-butyrolactone Conc. nitric acid Dimethyl formamide (hot)	Meta cresol 80% formic acid
Cellulose diacetate	70% (v/v) acetone Glacial acetic acid	Dichloromethane
Cellulose triacetate	Dichloromethane Glacial acetic acid	70% (v/v) acetone
Chlorofibre	Cyclohexanone Tetrahydrofuran Boiling dimethyl formamide	Cuprammonium hydroxide Conc. nitric acid
Cotton	Cuprammonium hydroxide	80% formic acid Xylene (boiling)
Flax and jute	75% (m/m) sulphuric acid (slowly and may be partly only) Swells in cuprammonium hydroxide	Xylene (boiling) Meta cresol
Modacrylic	γ-Butyrolactone Boiling dimethyl formamide	Cyclohexanone
Nylon	80% formic acid Cold meta cresol	Cuprammonium hydroxide Dichloromethane
Polyester	Hot meta cresol Dichloroacetic acid	Cold meta cresol 80% formic acid
Polypropylene	Boiling xylene	Hot and cold meta cresol 80% formic acid
Silk	Cuprammonium hydroxide Molar sodium hypochlorite	Meta cresol Xylene (boiling)
Viscose (including modal)	Cuprammonium hydroxide	Xylene (boiling) Boiling 5% sodium hydroxide
Wool and animal hairs	Molar sodium hypochlorite Boiling 5% sodium hydroxide	Cold meta cresol 80% formic acid

This table is intended as a general guide only, and is not exhaustive. The solvents named apply to normal examples of the fibre types listed. They do not necessarily extend to speciality, chemically modified or damaged fibres. Where mixtures are found, a fresh specimen should be used for each further test.

Table 2.2:   Melting points of synthetic fibres

Fibre type	Melting point (°C)
Acrylic	Does not melt – decomposes with discoloration
Aramid	Does not melt
Cellulose diacetate	250–255
Cellulose triacetate	290–300
Chlorofibre	185–210
Modacrylic	Does not melt – decomposes with discoloration
Nylon 11	182–186
Nylon 6	210–216
Nylon 6.6	252–260
Polyester	250–260
Polyethylene	133
Polypropylene	160–165

2.7 Melting points

A useful method for confirming the identity of synthetic polymer fibres is to determine the temperature at which they melt. This does not apply to natural fibres, which do not have thermoplastic properties, but can be a valuable means of establishing synthetic fibre types if brightfield microscopy and solubility tests have not been conclusive.

The most effective way to determine melting points is to use a hot stage, i.e. an addition to the microscope stage which is capable of being heated at a controlled rate and upon which the sample to be identified is mounted and observed. Several different types of hot stage are available, and each will give slightly different results. This is largely due to the method of mounting the fibre test specimen – some require the specimens to be inserted into a narrow glass tube, others heat the specimen sandwiched between two coverslips, while, in one particular method, the fibres are immersed in a low-volatility silicone fluid. The aim of all mounting methods is to ensure uniform conduction of heat to the specimen under test.

In operation, once the specimen has been mounted appropriately, it is located with the microscope and observed constantly as the device is switched on and the temperature begins to increase. The temperatures at which contraction, softening and melting take place are recorded. It is usual to carry out a preliminary 'sighting' test followed by three more accurate determinations in which the temperature is increased very slowly near to the values where change was recorded in the first test. If the specimen contains a mixture of fibres (as confirmed by fibres melting at different temperatures) each type will have to be observed separately.

Table 2.2 gives some typical melting point values, but it is important that the analyst should first conduct calibration tests on his own apparatus using authentic samples.

2.8 Staining tests

Although once a very popular method of fibre identification, the increasing use of blends and coloration techniques has made staining tests of only limited value today. Nevertheless, where an undyed specimen of fibre is the sample in question, staining tests can provide a rapid and simple indication of type. The presence of a mixture of fibres may be confirmed by examining a stained specimen microscopically, although it is difficult to compare colours produced under transmitted tungsten illumination with those that have been described and recorded under reflected daylight, i.e. as presented in the manufacturer's literature.

The best known textile fibre stains are the proprietary brands of Shirlastain. These are mixtures of dyes, each of which has an affinity for a particular fibre type. In use, the sample is immersed in the stain solution for a specified time, rinsed and dried, and its colour then read off from a chart supplied by the manufacturers (Shirley Developments Ltd, Manchester, UK).

Other fibre-specific stains are available, as are recipes for distinguishing between different variants of the same fibre type; however, their preparation and use are outside the scope of this book.

2.9 Refractive index measurements

Textile fibres are optically anisotropic, and have two principal RIs – one perpendicular to the fibre axis and one parallel to it. If a specimen of fibre is undyed or only lightly coloured, the determination of its two RIs (and the difference between them, known as the birefringence) can be a useful identification characteristic.

The simplest way of measuring the RI of a fibre is to mount samples in a series of different liquids of known and increasing RI until the sample becomes invisible. This occurs when the index of the fibre matches that of the liquid. In practice, to obtain accurate measurements, plane-polarized monochromatic light is used to illuminate the fibre and a range of liquids are used as mountants until those which cause the fibre to become invisible, (a) for light polarized in a plane along the axis and (b) for light polarized in a plane perpendicular to the axis, are found. The two values of the liquids are recorded and the birefringence may be calculated by subtracting $n\perp$ (perpendicular index) from $n\|$ (parallel index). *Table 2.3* shows the RIs of some common mounting media, while *Table 2.4* shows the values of $n\|$ and $n\perp$, and the birefringence of some common fibre types.

Table 2.3: RI of mounting media

Mountant	RI (nD^{20})
Water	1.33
n-Heptane	1.39
n-Decane	1.41
Glycerine jelly	1.45
Liquid paraffin	1.47
Olive oil	1.48
Cedarwood oil	1.513–1.519
Canada balsam	1.53
Methyl salicylate	1.537
o-Dichlorobenzene	1.549
Bromobenzene	1.560
1-Bromonaphthalene	1.658
Di-iodomethane	1.74

Table 2.4: RI of fibres

Fibre	$n\|$	$n\perp$	Birefringence
Acrylic	1.511	1.514	−0.003
Cellulose diacetate	1.477	1.472	0.005
Cellulose triacetate	1.469	1.469	0
Chlorofibre	1.541	1.536	0.005
Cotton	1.577	1.529	0.048
Flax	1.590	1.525	0.065
Modacrylic	1.520	1.516	0.004
Nylon 6	1.575	1.526	0.049
Nylon 6.6	1.578	1.522	0.056
Polyester	1.706	1.546	0.160
Polypropylene	1.530	1.496	0.034
Silk	1.591	1.538	0.053
Viscose (ordinary)	1.542	1.520	0.022
Wool	1.557	1.547	0.010

The values given are for guidance only and variations on these figures may be found in some cases.

2.10 Polarized light microscopy

The anisotropic nature of textile fibres referred to in Section 2.9 imparts characteristic behaviour when the fibres are examined under a polarizing microscope, i.e. between crossed polars. This is a very useful identification technique and can be applied to all but the most deeply dyed fibres. Because of the complexity and range of information obtainable by using polarizing filters, however, the subject of polarized light microscopy of fibres is considered separately in Chapter 4.

2.11 The identification of natural fibres

While the chemical and physical tests described in the preceding sections will allow the microscopist to distinguish between natural and man-made fibres, and to recognize specific types of regenerated and synthetic fibres, they will not be of great help in identifying different natural fibres. This is because natural fibres have only two basic chemical forms – cellulose for vegetable fibres and the protein keratin for animal fibres. Instead of chemical and physical tests therefore, natural fibres must be distinguished by their surface and interior features as seen under the microscope, usually with ordinary brightfield illumination. It is not the purpose of this book to offer a comprehensive treatise on the identification of natural fibres, which is a specialized subject requiring considerable experience (although the special case of the analysis of animal fibre blends is considered in more depth in Chapter 9). To provide the reader with some guidelines, however, the following notes should be helpful.

2.11.1 Vegetable fibres

Examine the fibres along their length, noting the size of their component cells (ultimates) and whether these are separate or in bundles. Transverse lines and their form should be noted. If possible, determine the average length of the ultimate cells. Single fibres showing convolutions, and where at higher magnifications diagonal (spiral) lines may be seen, will be cotton. Fibres which have bundles of ultimates will be bast, leaf or fruit fibres, e.g. flax, sisal or coir. Cross-sections are of great value in confirming the identity of vegetable fibres, and it is strongly recommended that such sections are prepared (see Chapter 5).

2.11.2 Animal fibres

With the exception of silk, all animal fibres are covered by a scaly outer layer called the cuticle. The shape, size and prominence of the scales in this cuticle offer perhaps the single most characteristic feature of animal fibres. Angular prominent scales suggest wool, while smoother scale edges are shown by the speciality fibres. Cashmere undercoat, in particular, has scales which are longer along the axis than most other fibres, and this coupled with the fineness and distribution of any pigment is a distinguishing feature. Some animal fibres have a medulla, a central core of air-filled cavities, and this can be another identifying characteristic. Fur fibres have ladder-type medullae. Cross-sections, particularly for silks, are another very valuable way of identifying animal fibres. As stated earlier, comparison of any unknown fibre with a series of authentic reference samples is always desirable, and this is particularly true with natural fibres.

For further information on fibre identification, the reader is advised to consult one of the standard reference works on the subject. The book by The Textile Institute (1975) provides a comprehensive guide to the identification of all types of fibres; Catling and Grayson (1982) have produced a detailed manual on the identification of vegetable fibres; and an atlas of the principal characteristics of animal fibres is provided by Appleyard (1978).

Note: asbestos fibres

Asbestoses are naturally occurring fibrous mineral silicates of which there are three principal types: chrysotile, amosite and crocidolite. The diameters of individual asbestos fibrils are orders of magnitude smaller than those of other textile fibres and this, together with their inorganic chemical nature and well publicized health hazards, makes the identification of these fibres a specialist subject in its own right.

Among the techniques used for identifying specific types of asbestoses are X-ray diffraction, transmission electron microscopy (TEM), infrared (IR) spectroscopy, differential thermal analysis and polarized light microscopy – the latter particularly when using the McCrone dispersion staining objective.

For further information about the properties and identification of asbestos the reader is advised to consult a specialized work on the subject, e.g. that by McCrone (1987) or the Health and Safety Executive of the UK (1995).

References

The Textile Institute. (1975) *Guide to the Identification of Textile Materials*, 7th edn. The Textile Institute, Manchester.

Appleyard HM. (1978) *Guide to the Identification of Animal Fibres*, WIRA (now British Textile Technology Group), Leeds.

British Standard BS 4658. (1978) Methods of test for textiles – preparation of laboratory test samples and test specimens for chemical testing.

Catling D, Grayson J. (1982) *Identification of Vegetable Fibres*. Chapman & Hall, London.

Health and Safety Executive, UK. (1995) *Methods for the Determination of Hazardous Substances*, No. 77. HMSO, London (in press).

International Standard ISO 5089. (1977) Equivalent to BS 4658 above.

McCrone WC. (1987) *The Asbestos Index*, 2nd edn. McCrone Research Institute, Chicago, IL.

3   Fibre Measurement

One of the most important properties of a textile fibre is its thickness or diameter. The diameter of a fibre affects many factors, a major one being the fineness of yarn that can be spun from it. Because it is necessary to have a certain number of fibres in a cross-section of yarn to ensure inter-fibre cohesion with staple (cut or finite length) fibres, the finer the fibre the finer the yarn that can be spun from it. This in turn will govern the end use of the fibre, as fine and coarse yarns are used in different types of products. In man-made continuous filament yarns, the diameter or fineness of individual filaments will have a great effect on the properties of the yarn, influencing such things as bulk, handle, flexibility and appearance. The importance of fibre fineness to the textile industry may be emphasized by the fact that fibres are usually specified as a diameter followed by the fibre type, e.g. 6 denier nylon; 21 micron wool.

3.1 Terminology

The diversity of the textile industry, both historically and in terms of fibre type, has meant that a number of expressions have evolved to describe the fineness of fibres, and the majority of these remain in use today. Some are based on subjective assessments but are supported by objective measurements, while others express the fibre fineness in terms of linear density. The latter are the only accurate means of comparing non-circular man-made filaments. *Table 3.1* shows some of the terms used to specify the fineness of fibres, and their meanings.

3.2 The measurement of fibre diameter

It may be seen from *Table 3.1* that many methods of expressing the fineness of fibres are indirect, i.e. they do not specify actual diameters but are based instead on mass per unit length measurements. Some other methods do specify mean diameters but the values are obtained from related properties, e.g. the air permeability of a standard mass of fibres. This latter is the case for the 'micronaire' test for cotton fibres, and the 'airflow' test for wool.

Table 3.1: Expressions of fibre fineness

Term	Meaning	Applied to:
Decitex	Decigrams per 1000 m of fibre	Man-made fibres, filaments
Denier	Grams per 9000 m of fibre	Man-made fibres, filaments
Micron	Fibre diameter in micrometres	Wool and animal fibres
Micronaire	Air permeability of standard plug of fibres	Cotton
Tex	Grams per 1000 m of fibre	Man-made fibres, filaments
Quality	Originally based on spinning possibility – now cross-matched to micron value	Wool and animal fibres

For many purposes, it is necessary to know not just the actual diameter of a fibre in microns but also the mean and individual values of a group or sample of fibres, and the spread of their distribution. The 'diameter distribution' of a sample of fibres will allow the experienced technologist to assess the spinning capabilities, yarn characteristics, physical behaviour and possible origins of the fibre. It can also provide an unambiguous comparison of suspected batch variations and indicate whether a sample is of a single fibre type or has been formed by the blending of one or more different fibres of the same generic type.

3.3 Standard method for the measurement of fibre diameter

Many procedures have been proposed for determining the fibre diameter distributions of circular or near circular textile fibres microscopically but, to date, only one method has gained universal acceptance. This method, specified in British and International Standards (BS 2043 (1968), ISO R137 (1975), IWTO 8-89(E) (1989)), is based on the use of a projection microscope and is outlined below.

3.3.1 Principle

Magnified images of short pieces of fibre are projected on to a screen. The widths of these fibres are measured using a graduated scale, and a special technique is used to avoid operator-induced bias. The arithmetic mean and coefficient of variation of a specified number of measurements are calculated and expressed in microns (micrometres) and percentages, respectively.

3.3.2 Apparatus

Projection microscope. The microscope must comprise a light source, condenser, X–Y stage, objective, eyepiece and a circular screen on to which the images of fibres are projected. The objective and eyepiece must provide a screen magnification of 500× (no further specifications for the optical components of the microscope are quoted).

The circular screen, which may be transparent or opaque (e.g. ground glass or white paper), must be capable of being rotated in its own plane about its centre, and must either have a scale calibrated in millimetres across its diameter or be equipped with a movable scale. A circle equal to one quarter of the distance from eyepiece to screen is marked in the centre of the screen, and all measurements are made inside this circle.

Microtome. Although referred to as a microtome in all published standards, this is really a device for preparing fibre snippets of a predetermined length rather than a microtome in the conventional sense. Various forms of the device have been made, but perhaps the most widely used in the textile industry is that commonly referred to as the *heavy duty Hardy microtome*, or its derivative approved by the International Wool Textile Organization, the *IWTO fibre microtome*. Both of these consist of an open slotted plate into which a bundle of several hundred parallelized fibres are placed. The slot is then closed with a second, tongued plate, and the protruding fibres are sliced off flush with each side of the plate. A key-type or threaded plunger is inserted into the base of the slot, and the held, parallel fibres are advanced proud of the upper surface of the plate by 0.4, 0.6 or 0.8 mm depending on the expected mean diameter of the fibre sample. The protruding tuft of fibres is sliced off with a scalpel or razor blade, and the resulting snippets form the sample for measurement.

Slides. As it will be necessary to examine each fibre in the sample separately, large slides and coverslips are used. The slide size specified is 75 × 40 mm, for which the suitable coverslip is 50 × 35 mm. A thickness of 0.13 – 0.17 mm (No. 1) is listed for the coverslips, but in practice they should be chosen to match the correction of the objective in use.

Mounting medium. Liquid paraffin is recommended as a mountant. The properties of this medium are described in Section 2.2.

3.3.3 Sampling and preparation

Due to the inherent variability of most textile materials, it is important that a representative sample of fibres is obtained for testing. The principles of sampling are outlined in Section 2.1, while specific procedures for fibre diameter measurement are given in the relevant standards (BS 2043, ISO R137, IWTO 8-89(E)). The diameter of many textile fibres varies according to temperature and humidity, and all samples for diameter measurement must be conditioned and prepared in the ISO standard atmosphere for textile testing of $65 \pm 2\%$ relative humidity and $20 \pm 2°C$.

3.3.4 Mounting

The fibre snippets sliced from the microtome are placed in a few drops of mounting medium on a microscope slide. The snippets are stirred well into the medium using a dissecting needle, so that an even distribution on the slide is obtained. A soft cotton cloth is used to wipe away sufficient of the mixture so that none will be squeezed out from under the coverslip. This avoids preferential loss of fine fibres. The coverslip is lowered on to the mixture from one side in the conventional way. It is usual to seal the edges of the coverslip, often with ordinary nail varnish, to secure it.

3.3.5 Calibration

Before use, the magnification of the microscope is confirmed and adjusted if necessary. A stage micrometer of 1 mm divided into 0.01 mm divisions is suitable. 0.1 mm of the micrometer should measure 50 mm on the screen when measured with the millimetre rule.

3.3.6 Measurement

Each part of the measurement procedure specified is designed to ensure two things: firstly, that the measurements are accurately made; secondly, that a random sample of the fibres present on the slide is measured.

The slide is placed on the stage of the microscope, coverslip towards the objective, and a corner of the coverslip is located. The slide is moved 'in' approximately 0.5 mm in the X and 0.5 mm in the Y direction, and the first field of fibres is brought into view and focus. Each fibre present in the field of view is measured in width, except:

(i)   images with more than half their width outside the central circle;
(ii)   images shorter in length than 5 cm;
(iii)   images which cross over another image.

Measurement is made by aligning the movable centimetre scale perpendicular to the fibre axis and positioning it so that a centimetre division coincides with one edge of the fibre. Correct focusing is achieved by adjusting so that the edge of the fibre at the centimetre division is sharp, while the other edge of the fibre shows a *white* Becke line. The number of millimetres from the sharp edge to the inside of the white Becke line is then recorded as the measurement of the width or diameter of the fibre. If, as often happens, the second edge of the fibre inside the white Becke line does not coincide exactly with a millimetre division it is recorded under the lower whole number of millimetres. A correction is then made for this in the final calculation.

When the fibres which qualify in this first field (if any) have been measured, the slide is traversed a distance greater than the longest fibre length in the X direction. Alternatively, it can be moved a distance of 0.5 mm in the X direction. Either of these steps will ensure that the chance of measuring the same fibre piece twice is minimal. The fibres present in this new field are focused and measured, the slide moved along in the X direction again, and the process repeated. At the end of this X traverse (when the other edge of the cover slip is reached) the number of fibres measured from the traverse is noted, and the number of traverses needed to obtain the desired number of measurements is calculated. The slide must be moved in the Y direction a minimum of 0.5 mm before the next traverse is made, again to avoid the possibility of the same fibre being measured twice. If the number of traverses needed cannot be accommodated in 0.5 mm steps in the X direction on one slide, it will be necessary to prepare a second slide.

Traversing and measurement are continued until the whole slide has been covered. The form of the traverse path is that of a raster as shown in *Figure 3.1.*

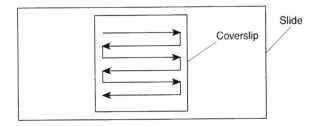

Figure 3.1: Path taken across slide during measurement.

3.3.7 Number of measurements

The number of measurements necessary to obtain a given degree of accuracy will depend on the spread of the diameter distribution. Details of the statistical calculations are given in the appropriate standards (BS

Table 3.2: Confidence limits for different numbers of measurements

95% confidence limits (%)	Number of measurements
± 1	2500
± 2	600
± 3	300
± 5	100

2043, ISO R137, IWTO 8-89(E)); however, *Table 3.2* offers a rough guide to the 95% confidence limits expressed as a percentage of the arithmetic mean for different numbers of measurements.

3.3.8 Calculation and expression of results

As the microscope magnification is 500×, the measurements in mm (lower limit + 0.5 mm) multiplied by two give the fibre diameter in micrometres (microns). A more popular method is to calculate the mean and standard deviation of the measurements directly, then to double the final figures. A value of one is added to the mean diameter as the focusing correction. It is also necessary to calculate the coefficient of variation of the measurements.

3.4 Projection microscopes

Two main types of projection microscope exist: those which project the image on to a translucent ground glass screen from behind and those which project the image on to an opaque (usually matt white) surface. There is a general consensus within the textile industry that the latter type is more comfortable to work with and causes less fatigue. Focusing an image on a solid white surface certainly seems easier than doing so on a screen, which is also projecting background light at the eye and which has the textured effect of ground glass.

In the past, high-quality, wide-field, inverted-projection microscopes were made specifically for the textile industry, notable examples being those produced by Bausch & Lomb and Vickers Instruments. Few examples of these remain, however, and the fibre technologist is now left with the choice of ground-glass screen projectors or the basic, narrow-field projection microscope made by a specialist company (R&B Instruments Ltd, Leeds, UK) for the wool industry. This microscope is essentially unchanged in design from the draft specification issued by the IWTO in the 1940s. Many hold the view that it suffers from a small projected field, low power illumination and inconveniently small size when compared with some of the instruments of the past, despite conforming with all the IWTO specifications. Those laboratories fortunate enough to retain versions of the Bausch & Lomb, Vickers or even early Zeiss projection microscopes made for fibre measurement are unlikely to dispose of them.

3.5 Alternatives to the standard method of fibre diameter measurement

It will be appreciated from the preceding sections that the determination of fibre diameter distribution by manual operation of a projection microscope is a far from ideal test method. The procedure is tedious and lengthy, intense operator concentration is required, measurement may be affected by subjective factors (the operator) and slide preparation is awkward. Not surprisingly, many attempts have been made over the years to devise an improved method of determining the diameter distribution of fibres – mainly to speed up the process. Until recently, none of these were considered viable alternatives to the standard method, but now two techniques are gaining increased acceptance.

The fibre diameter analyser (FDA) is a non-microscopical system, and operates by light scattering. Many thousands of fibre snippets are circulated in a flow of solvent through which a laser beam passes, and the degree of scatter of the laser light is recorded by sensors and translated into a diameter distribution. The makers of the system claim increased accuracy over the conventional method due to the greater sample size – many thousands of readings are taken over a relatively short time – but there have been difficulties in obtaining direct correlation with results from projection microscopy.

The optical fibre diameter analyser (OFDA) is a microscope-based system which uses image analysis to record the diameters of several hundred fibres as they are scanned automatically on a conventional compound microscope coupled to a CCD camera. Again, increased accuracy is claimed due to the larger sample size of this system but, as with the FDA, direct correlation with the standard method has sometimes proved elusive. It is expected, nevertheless, that ongoing work with the FDA and OFDA will soon see their use accepted by the standards organizations, and the projection microscope will gradually be replaced by these faster and less tedious alternatives. The cost of the new systems is likely to be their only constraint.

Another version of the (non-microscopical) light scattering method of fibre diameter measurement has been developed by the Commonwealth Scientific and Industrial Research Organization in Australia. The Sirolan Laserscan again uses the principle of laser light scattering to produce distribution, mean and standard deviation measurements of wool fibre diameter, and has also been proposed for acceptance as a standard method by the IWTO. Meanwhile, on the microscopical side, the development of a computer algorithm to correct for focusing effects, to allow automated measurement of the images of fibres without accurate focus adjustment, has been proposed for the measurement of the diameter of cashmere fibres. It is intended to incorporate the algorithm into an image analysis system, which would overcome the need for constant refocusing when measuring fibre diameter by image analysis.

References

British Standard BS 2043. (1968 (1993)) Wool fibre fineness by projection microscope.
International Standard ISO R137. (1975) Determination of fibre diameter – projection microscope method.
International Wool Textile Organization (IWTO) 8-89(E). (1989) Method for determining wool fibre diameter by projection microscope.

4 Polarized Light Microscopy

The main advantage of polarized light microscopy is that it allows identification of synthetic fibres which have an otherwise uniform appearance under the light microscope. For this reason, a fibre microscopist would be well advised to make sure when obtaining a light microscope that it is capable of using polarized light to examine specimens. The reason for the ability of polarized light microscopy to identify fibres is that it can give information about the orientation of the long chain molecules that make up a fibre. This information can be interpreted either in a semi-quantitative manner in order to identify fibres or quantitatively to measure the degree of orientation in a fibre of known type.

4.1 The microscope

A light microscope which is suitable for examination of fibres using polarized light requires as a minimum two devices which are capable of producing plane-polarized light to be placed in the optical pathway. One, known as the polarizer, is placed between the light source and the sample, which is then as a consequence illuminated with plane-polarized light, and a further one, known as the analyser, is placed after the sample. These devices are collectively known as polars and are usually made of 'Polaroid'. It is possible to place pieces of sheet Polaroid in the appropriate locations of an ordinary light microscope to give some simple capabilities with polarized light.

Purpose-built polarizing microscopes usually have a number of extra features not found on a standard light microscope. Some of these are described below.

4.1.1 Removable polarizer and analyser

The ability to remove the polarizer and analyser means that the specimen can easily be examined either in ordinary light, in polarized light with just the polarizer present or between crossed polars with both polarizer and analyser present. It is important that at least one of the polars can be rotated so that the crossed polar position can be set exactly at extinction. A rotating analyser graduated in degrees is needed for the Sénarmont method of measuring orientation.

4.1.2 Rotating stage

It is particularly important that the stage of the microscope can rotate as the appearance of the sample between crossed polars varies with its orientation with respect to the plane of polarization. Examination of samples and measurements are usually made with the specimen oriented at 45° to the plane of polarization of both the polarizer and analyser.

4.1.3 Body slot

This is placed in the body of the microscope between the objective lens and the analyser in order to allow the insertion of either a fixed or a variable birefringence plate into the optical path. The body slot is arranged at 45° to the plane of polarization of both polarizer and analyser so that the inserted waveplate either adds to or subtracts from the sample birefringence. Fixed value waveplates are useful for semi-quantitative analysis whereas variable waveplates, known as compensators, are necessary for quantitative measurement.

4.1.4 Monochromatic light

This is usually produced by means of a filter inserted in the light path before the condenser lens. It is required for quantitative measurements as the optical path difference necessary to produce a whole wavelength retardation varies with the wavelength of the light used. A more powerful lamp for the microscope is generally required if a filter is used to produce monochromatic light.

4.2 Birefringence and optical path difference

In order to understand polarized light microscopy, it is necessary to appreciate the role that the RI of the material plays in producing the final image in this type of microscopy. Most fibres have an RI for light vibrating parallel to the fibre length which is different to that for light vibrating perpendicular to the fibre length. This difference is known as the birefringence and it is this which can be revealed by the use of polarized light.

The RI of a material is a measure of how much light is slowed down as it passes through it; the higher the index the slower is the passage of the light wave. It is dependent on the amount of covalent bonding in the chemical structure of the material which exists in the same plane as the light vibration direction. In a crystalline material there can be up to three separate refractive indices, one for each of the three axes of the material. These are often designated $n\gamma$, $n\beta$, and $n\alpha$, where $n\gamma$ is the highest RI and $n\alpha$ is the lowest. The highest and lowest RI directions are sometimes also known as the slow and fast directions, respectively. If the molecular structure of the material is the same in all directions, these RIs will all have the same value and the material is said to be isotropic. If the structure is different in different directions then the values of the three RIs will not necessarily be the same and the material is said to be anisotropic.

In most fibres, the chain molecules are oriented approximately parallel to the fibre direction but are randomly arranged perpendicular to it so that the optical properties can generally be described by just two RIs, one parallel to the fibre axis and one perpendicular to it, designated $n\|$ and $n\perp$, respectively. If the fibre has a more ordered structure within its cross-section, such as the molecules being grouped into sheets or a structural difference between sheath and core, then the perpendicular RI may vary across the fibre.

The value of the RI parallel to the fibre axis $n\|$ is dependent on the RI of the polymer crystal parallel to the polymer chain axis, and in most fibres it usually has the highest value and is equivalent to $n\gamma$ for the polymer crystal. The perpendicular RI $n\perp$ is an average value of the other two RIs of the polymer crystal $n\alpha$ and $n\beta$. In the molten polymer, because the chains are randomly oriented, the overall RI is an average of all three RIs and is the same in all directions, i.e. the material is isotropic.

It is possible to determine each of the two RIs of a fibre directly using the Becke line test, the Becke line being the bright fringe which appears on the boundary of the specimen when it is slightly out of focus. If the fibre is examined in the microscope with just the polarizer in place (i.e. with the analyser removed), it will be illuminated with plane-polarized light. If the fibre is then rotated until it is parallel to the vibration direction of the light, the RI parallel to the fibre axis becomes the one that is active. If the fibre is mounted in a liquid whose RI matches the $n\|$ of the fibre, then the fibre edge will become invisible. Similarly, if the fibre is rotated through 90° the RI perpendicular to the fibre axis, $n\perp$, becomes the dominant one. If a mounting liquid of known RI can be found to match the RI of the fibre in each of these two cases, the two indices of the fibre will have been determined.

The Becke line test helps in this task by determining whether the mounting liquid has a higher or lower RI than that of the fibre. When the microscope focus is adjusted so that the objective lens is moving away from the specimen, the Becke line moves towards the material with the higher RI. The strength of the Becke line also varies directly with the difference in the RI. Therefore, with a judicious choice of mounting liquid with reference to possible values of the fibre RI, each RI of the fibre can be determined with the minimum number of changes of mounting liquid. The Becke line effect is heightened by stopping down the condenser aperture.

The difference between the two values of RI is known as the birefringence and it is defined as $(n\| - n\perp)$ which means that it can take both positive and negative values. For most fibres the value is positive, acrylic being the most notable exception with a negative birefringence. This reflects the fact that the value of $n\|$ is generally higher than that of $n\perp$. The values for the individual RIs and for the birefringence of most common fibres are shown in *Table 2.4*.

The birefringence of a fibre depends both on the polymer type and on the degree of orientation of the polymer chains such that determination of the fibre birefringence can be used either to identify the fibre type or to measure the orientation of a fibre whose type is known. The ability to identify fibres is reliant on the fact that the orientation of commercial fibres generally falls within well-defined limits.

When a fibre is viewed in the microscope between crossed polars, the effect that is seen depends on the optical path difference (OPD) which is the product of the birefringence and the specimen thickness.

$$OPD = \text{Thickness} \times (n\| - n\perp).$$

In a whole fibre, because it is roughly circular, the thickness varies from the edge to the centre, so that it is the centre of the fibre which is examined. In order to determine the birefringence of a specimen it is therefore necessary to know its thickness.

4.2.1 Examination in the microscope

When the polarizer and analyser of the microscope are crossed and there is no sample present the field of view appears completely black. If a birefringent sample is placed between the crossed polars and rotated, it will appear to change from bright to almost black four times in a full 360° rotation. An isotropic specimen under the same conditions will appear black at all times.

When the polarized light from the polarizer encounters a birefringent specimen, it is split into two components, one of which travels through the material at a speed determined by $n\|$ and one which travels through it at a speed determined by $n\perp$. The difference in speed of the two waves taken together with the distance they travel through the specimen means that there is an OPD between the two waves which results in a phase difference. At the analyser, the components of these two waves which are vibrating parallel to the analyser direction recombine and thus can interfere with each other according to the phase difference. With monochromatic light, OPDs of 0, λ, 2λ.....nλ (where λ is the wavelength of the light in use) result in destructive interference, i.e. the specimen appears black.

If the specimen is oriented parallel or perpendicular to the direction of polarization of the polarizer then only one beam passes through the specimen but, as this beam is vibrating parallel to the polarizer direction, it is blocked by the analyser. This is why the specimen appears black in these positions during rotation. The maximum brightness is produced at the intermediate diagonal positions at 45° to the polarizer and analyser. Observations and measurements are made with the specimen in one of these positions.

In monochromatic light, when only one wavelength of light is involved, the specimen also appears black between crossed polars whenever the OPD is a whole number of wavelengths. This means that a birefringent material cut in the form of a wedge and viewed between crossed polars will appear to have a series of equally spaced black lines where this condition is fulfilled.

In white light, which is used for most observations, the situation is more complicated because the wavelengths which make up the white light vary from violet λ = 400 nm to red λ = 700 nm. Each of these wavelengths interferes destructively at the OPD corresponding to the wavelength so there is no single OPD where they are all absent. However, because of the loss through interference of particular wavelengths, the resulting light is coloured by those wavelengths which remain. In a wedge-shaped sample of increasing OPD in white light there is a distinct sequence of interference colours which fall into approximate bands corresponding to λ, 2λ, 3λ etc. (the wavelength of 551 nm is taken as the average value). These colour sequences are referred to as first order, second order, third order etc. The colours in the first order are quite strong and can be easily recognized but, as the order increases, the colours become more washed out until, at high orders, they become a uniform white. Colour printed charts of the sequence of interference colours known as Michel-Lévy charts are available, usually from the microscope manufacturers.

4.3 Identification of fibres

The use of polarized light microscopy for fibre identification depends on visually estimating the OPD by identifying the colour that the sample shows on the interference colour chart. It is most useful for the man-made fibres which otherwise appear featureless when examined in the microscope. The identification rests on the assumption that commercial fibres of a particular polymer type always have approximately the same amount of orientation. If the fibre is assumed to be circular in cross-section, measurement of its width is equivalent to the specimen thickness and hence its birefringence can be calculated. In practice, many commercial fibres have diameters which fall into a narrow range so that they can be identified purely by comparative appearance. Ideally, the unknown sample should be compared to a known sample of a similar diameter.

To identify an unknown fibre, it should be examined in the diagonal position between crossed polars. The colour that the specimen appears is then identified on the Michel-Lévy chart. The low order colours are easiest to identify whereas the high order colours shown by fibres such as nylon and polyester are harder to place but are in themselves a pointer to identification.

To assist with identification, plates of known OPDs can be inserted into the body slot of the microscope. The two most common plates in use are quarter wavelength $(1/4\lambda)$ (again, the average wavelength is taken to be 551 nm) and whole wavelength (1λ). When two birefringent objects are placed between crossed polars so that they are both in the optical path, they can be oriented with respect to one another so that their individual OPDs either add or subtract. If the two objects are placed so that their directions of highest RI (i.e. slow directions) are parallel to one another then the two OPDs are added together. Conversely, if the direction of lowest RI (fast direction) of one object is placed perpendicular to the direction of highest RI (slow direction) of the other then the two OPDs subtract from one another.

The fixed-wavelength plates have the slow direction marked on them so that during observation the unknown fibre can be rotated such that its axis is first parallel to and then perpendicular to the slow direction of the fixed waveplate. If the fibre has a positive birefringence, which most do, the interference colours will be displaced up the colour scale when the slow direction of the waveplate is parallel to the fibre axis and down the colour scale when the slow direction of the waveplate is perpendicular to the fibre axis. In the case of a fibre with negative birefringence, the effect will be the opposite to the above so that the sign of the birefringence can be determined from the change in direction of the colour. The colour shift caused by the $1/4\lambda$ and 1λ waveplates is also used to determine precisely where on the colour chart the unknown fibre lies. In particular, the plates are used to differentiate between first order white and higher order whites as the latter are hardly affected by a small increase or decrease in OPD.

These simple tests, if used in conjunction with fibre samples of known origin, are usually enough to identify the most common commercial man-made fibres that have few other microscopical features. The fibres which generally cause the most difficulty are those with high birefringences such as polyester and nylon.

4.4 Quantitative aspects

The measurement of OPD depends on the effect which has been described in the previous section, that is the subtraction of the OPDs of two birefringent objects which have been placed with their slow directions at right angles to one another. If the OPD of the sample under test is opposed by an OPD which is exactly equal to it then the two will cancel each other out completely and leave the sample appearing black. This effect is known as compensation, and the device which is used to oppose the OPD of the sample is known as a compensator. If the OPD of the sample is known then measurement of its thickness, using the same units as are used for the OPD, allows its birefringence to be calculated.

Compensators are devices which have an OPD which can be varied in a controlled way so that it can be read from the device or calculated from the readings. There are a number of ways of obtaining this effect but the devices in use tend to fall into one of two groups. One type can measure small OPDs, of less than one wavelength, with great accuracy. The other type can measure up to a large number of orders but with less accuracy.

The simplest type to understand takes the form of a wedge cut from birefringent material, usually quartz, which increases in OPD because of its increase in thickness. Measurement of the distance of the point of extinction from the end of the wedge allows the OPD to be calculated.

The Sénarmont method allows OPDs of up to one wavelength to be measured accurately. A special, accurate $1/4\lambda$ waveplate is used whose slow direction is parallel to the polarizer when inserted into the body slot. The analyser is first rotated to give extinction in the required part of the specimen, the nearest extinction point to the zero setting being chosen. The specimen is then rotated through exactly 90° and the corresponding extinction position is found by rotating the analyser in the opposite direction from zero. Half the sum of the two angles is used to calculate the OPD from the following equation:

$$OPD = \frac{\text{Angle} \times \text{Wavelength } \lambda}{180} \text{ nm.}$$

The measurement needs to be carried out in monochromatic light. It is possible with this method to measure the fractional part of higher order path differences if the whole wavelengths can be determined by other means.

Higher order path differences are measured by the use of a tilting compensator in which a magnesium fluoride, quartz or calcite plate is tilted by means of a graduated scale to give a variable OPD. These compensators are usually made to cover different ranges, typically five wavelengths, 10 wavelengths and 30 wavelengths, depending on the type of plate used. As both the thickness and birefringence of the plate change as it is tilted, it is necessary for the OPD to be read from a table for the observed angle of tilt. It is usual to tilt the plate in both directions to obtain compensation on either side of zero and use half the sum of the tilting angles for the calculation of the OPD.

One problem which can arise when measuring path differences greater than one wavelength is due to the fact that compensation takes place at every whole wavelength path difference. Therefore, it is necessary to distinguish the true zero compensation from similar positions one or more whole wavelengths displaced from this position. In white light, only the true zero position should appear black but, due to the fact that in some materials the birefringence changes with wavelength, a property known as dispersion, this is not always the case. In order for the zero position to appear black, the dispersion of the compensator would have to match that of the sample. In fact, in the case of a quartz compensator paired with a polyester fibre, the first-order fringe appears black when the path difference of the fibre is six wavelengths. One way of overcoming this problem is to cut a wedge on the end of the fibre and count the number of fringes from the thin end of the wedge to the centre of the fibre, so giving the whole number of wavelengths. Another solution to the problem is to construct a compensator from the same material as the sample; for example, polyester film in the case of polyester fibres. This can be used to compensate a large part of the path difference leaving the rest to be measured by a conventional compensator.

4.4.1 Orientation

With a fibre of known type it is possible to measure the average orientation of the polymer chains with respect to the fibre axis by use of Hermans orientation factor f which is defined as

$$f = \frac{\text{Actual birefringence}}{\text{Theoretical maximum birefringence}}.$$

The theoretical maximum birefringence is calculated by assuming that $n\|$ is equal to $n\gamma$ for the fully oriented polymer and that $n\perp$ is equal to the average of $n\alpha$ and $n\beta$ for the polymer. The values of the RI for the oriented polymer can be obtained from the literature. The value of f would be 1 if all the polymer chains were oriented parallel to the fibre axis and 0 if there was no orientation.

The relationship between the orientation factor and the orientation of the polymer chains is given by the following relationship:

$$f = 1 - \frac{3}{2}\sin^2\theta.$$

For the purposes of this equation the chains are all assumed to have a uniform orientation of θ to the fibre axis.

Further information on the practical use of polarized light microscopy applied to polymers and fibres may be found in Hartshorne and Stuart (1970), Hemsley (1984, 1989) and Robinson and Bradbury (1984), while two Textile Institute publications (1975; Greaves, 1992) provide colour micrographs illustrating the appearance of different man-made fibres under polarized light.

References

The Textile Institute. (1975) *Guide to the Identification of Textile Materials*, 7th edn. The Textile Institute, Manchester.

Greaves PH. (1992) Microscopy, imaging and analysis. In *Advances in Fibre Science* (ed. SK Mukhopadhyay). The Textile Institute, Manchester, pp. 47–66.

Hartshorne N, Stuart A. (1970) *Crystals and the Polarising Microscope*, 4th edn. Arnold, London.

Hemsley DA. (1984) *The Light Microscopy of Synthetic Polymers*. RMS Handbook No.7, Oxford University Press, Oxford.

Hemsley DA. (1989) *Applied Polymer Light Microscopy*. Elsevier, London.

Robinson PC, Bradbury S. (1984) *Qualitative Polarized Light Microscopy*. RMS Handbook No. 9, Oxford University Press, Oxford.

5 Special Preparation Techniques for Light Microscopy

5.1 Cross-sections

There are four main techniques for preparing cross-sections of textile materials for light microscopy. For all but the most specialized examinations, two of these techniques, the plate method and the Hardy microtome, will provide all the information required by the microscopist. The other two methods, the mechanical microtome and the grinding procedure, are more suited to research laboratories or special investigations where time is not of primary importance.

5.1.1 The plate method

For routine examinations of fibre cross-sectional shape (contour), e.g. for identification, the speed and simplicity of the plate method is unrivalled. Essentially, the 'plate' consists of a hard steel microscope slide, through which holes approximately 0.75 mm in diameter have been drilled. The sample of fibres to be sectioned is pulled through one of the holes, the protruding ends cut flush with the smooth surfaces of the plate, and the resulting section (which will be the thickness of the plate) is examined with either transmitted or reflected light by placing the plate carefully on the microscope stage and locating the hole in which the fibre sections are held. Thus, the plate itself is used as the slide. The preparation sequence is shown diagramatically in *Figure 5.1*.

To increase contrast between the fibres and background in transmitted light examinations, black oil paint may be rubbed carefully over the section, and the excess removed with a clean cloth moistened with solvent. This produces a black background against which the fibres appear bright, transmitting light by internal reflection.

A disadvantage of the plate method is that the sections produced are necessarily the thickness of the plate, usually 0.5 mm. For studying fibre detail this is too great and, because it is also difficult to control all fibres equally when cutting, not all fibres will be perfectly flush with the plate. Use of the plate method, therefore, is normally confined to straightforward shape identification.

(a)

(b)

(c)

(d)

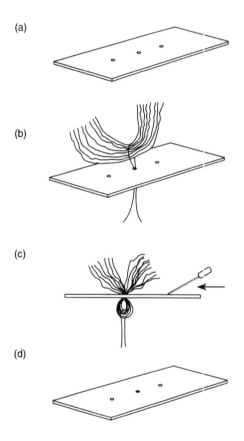

Figure 5.1:   Cross-sectioning by the plate method. (a) The plate. (b) The fibres to be sectioned are pulled through a hole in the plate with thin, strong thread. (c) When fully through the plate, the protruding ends of the fibre sample are sliced off flush with the surface of the plate using a safety razor or scalpel blade. (d) The hole now contains a cross-section of the fibre sample of the same thickness as the plate.

5.1.2 The Hardy microtome

Named after its inventor in the United States (Hardy, 1933), the Hardy microtome is well known in textile laboratories. Small enough to be hand held, the microtome is made up of two steel plates which slot together in a central tongue and groove. The fibre sample to be sectioned is placed in the groove of the device, and the tongued part is then pushed home to secure the fibres. A small amount of embedding medium, either 3% cellulose nitrate in 50/50 di-ethylether/ethanol or 6% cellulose acetate in acetone, may now be applied to the base of the tuft of fibres on either side of the microtome where it is held in the groove. When this medium has hardened, the protruding fibres are sliced off flush with the microtome on either side with a safety razor or scalpel blade. The pusher of the microtome, a threaded plunger the size of the fibre-filled groove, is then rotated into position and advanced by the amount required to push the fibres proud of the microtome's surface (usually 10–20 µm). A small amount of embedding medium, as used earlier, is applied to the exposed fibre ends. Once this has hardened, the fibre ends are sliced off flush and mounted on a microscope slide still embedded in the medium.

Hardy microtome sections are much thinner than plate ones, and can be used to study dye penetration, pigment distribution and medium-to-fine structural detail, as well as straightforward contour. The device is easy to use and with practice a satisfactory section can be obtained in 15–30 min. *Figure 5.2* shows a Hardy microtome and the way a bundle of fibres is held before slicing flush.

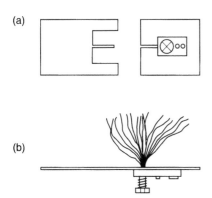

(a)

(b)

Figure 5.2: The Hardy microtome. (a) The microtome separated into two halves. (b) Side view showing fibres held in the slot prior to slicing flush the upper portion.

5.1.3 Mechanical microtome sections

The plate and Hardy microtome methods will suffice for 90% of textile studies. Where thinner sections are required, however, conventional precision microtomes may be used, whether of the rocking, sliding or rotary type. It is important that the specimen to be sectioned is no larger than 2–3 mm^2 or damage to the blade could result. Fibres are generally aligned in Beem-type capsules then embedded in Araldite or methacrylate resin. For soft fibres such as animal, wax embedding may also be used. Once the specimen is satisfactorily aligned and embedded, the microtome is used in the normal way.

5.1.4 The grinding method

The methods so far described are suitable for preparing fibre or yarn sections. When a larger specimen, e.g. a fabric, is required to be examined, then the grinding method must be used. In this, the specimen is mounted in a mould and embedded in Araldite or some other resin. Vacuum impregnation may be used if there are many air bubbles trapped in the structure. When the resin has set, the mould is removed and one face of the block is ground flat on silicon carbide papers of decreasing particle size. A 2–4-mm disc of this flattened block is cut out and the flat face carefully stuck to a microscope slide with a thin layer of resin. The upper (non-flat) face of the resin is then ground flat in the same manner as before, until the required thickness of section is reached. The resulting section, when washed clean, is examined in the normal way. The grinding method is particularly useful where spatial positions of yarns and fibres are to be maintained, but it is a relatively lengthy procedure.

5.2 Scale casts

The increasing popularity of speciality animal fibres and their high cost have led to a resurgence of interest in methods of positively identifying these fibres. As mentioned in Section 2.11, the distinguishing feature of animal hair fibres is the scaly outer cuticle layer. The nature of this scale pattern allows the specific type of animal fibre to be identified and enables a distinction to be made between wool, cashmere, camel hair, yak and mohair, for example. As cashmere is several times more expensive than wool, accurate identification is important. There is occasionally a misconception that cashmere can be identified by diameter measurement alone. This is incorrect, as many fine lambswool fibres can be as fine or finer than some de-haired cashmeres.

When the fibre to be identified is undyed and non-pigmented, a straightforward examination of the cuticle and comparison with reference samples will generally permit positive identification. Uniformity of diameter, scale length and height comparisons, and general diameter may serve as confirmatory features. If the fibre is heavily dyed, however (as most finished textile goods are), or is pigmented then the scale pattern will not be visible with transmitted light. Laboratories fortunate enough to possess a scanning electron microscope (SEM) face no problem here as the SEM will display the fibre surface with even greater clarity than the light microscope. There are many textile labs, however, which do not have an SEM yet require fairly rapid confirmation of the identity of an animal fibre. The simple way to achieve this is to make a replica of the surface of the fibre (called a scale cast) and examine it under the standard transmitted light compound microscope.

Casts may be made in many ways, a rough and ready one being to simply use cellophane adhesive tape. The quality of such casts though is generally poor and a far better way is to use a thermoplastic medium. In this method, a solution of polymer in solvent (polyvinyl acetate is most commonly used: a 20% solution in dichloromethane or other suitable solvent) is first spread evenly over a microscope slide. The solvent is allowed to evaporate, leaving a hard layer of polymer on the slide. The fibres to be identified are placed on this layer of polymer and another microscope slide placed over them. This composite slide is then transferred to a hotplate and a weight placed on top of it to provide suitable pressure. A second slide is positioned next to the weighted composite slide and a small piece of solid polymer (of the same type used to make the initial layer) is placed on the second slide to act as a temperature guide. The temperature of the hotplate is then gradually increased until the solid piece of polymer can be seen to soften, at which point the heat is turned off and the slides allowed to cool. Once cold, the weights are removed and the top slide separated from the bottom slide and embedded fibres. The fibres themselves are now removed, taking care not to damage the impression they have left in the polymer. The scale casts of the fibres may now be examined with transmitted light.

Another way of making scale casts is to use gelatin. A 3% solution in water is spread over a microscope slide and the fibres placed in it. When the water has evaporated, the fibres are removed from the hardened gelatin and the resulting casts examined as before. Gelatin casts tend to be less durable than thermoplastic polymer ones.

These methods will allow clear examination of the scale patterns of even the heaviest dyed or pigmented animal fibres.

Reference

Hardy JT. (1933) A practical laboratory method of making thin cross sections of fibres. U.S. Dept of Agric. Circ. No. 378.

6 Other Light Microscopical Techniques Applied to Fibres

6.1 Fluorescence microscopy

Certain compounds have the ability to absorb incident electromagnetic radiation (e.g. light) and transfer electrons to a higher energy level or 'excited' state. As the electrons are not stable in this excited state, they must return to their ground states but in doing so lose energy. The lost energy is emitted as light of longer wavelength than that which caused the initial excitation – generally visible light when the incident exciting light has been near ultraviolet (UV) (300–400 nm). This phenomenon of converting invisible UV radiation into longer wavelength visible light is called fluorescence, and compounds which have this property are called fluorochromes.

A fluorescence microscope uses a light source (often a mercury vapour lamp) to provide an intense broad band emission of radiation, typically from 300 to 700 nm. This light then passes through an exciter filter which transmits only the wavelengths to be applied to the specimen. The light which is emitted by the specimen (i.e. the fluorescence) passes through the microscope objective, a barrier filter to prevent the passage of any stray reflected UV light (and a chromatic beam splitter or dichroic mirror in the case of incident excitation), and on to form the primary and subsequent secondary images.

The information contained in an image from a fluorescence microscope is a combination of chemical and structural features of the specimen. Which parts fluoresce and which do not, at which excitation wavelengths and which emission wavelengths, and at what intensity of fluorescence can all be used to gain information about the specimen which may not be available under other illumination conditions.

There are two main reasons for applying fluorescence microscopy to fibres: (i) to study structurally the fibre itself and (ii) to establish the presence or absence of a substance on the fibre. Many natural and some synthetic fibres have a natural fluorescence at certain excitation wavelengths and this fluorescence (called autofluorescence) can be altered by chemical changes to the fibre. It is possible therefore to use fluorescence microscopy to investigate the past chemical history of a particular fibre. Examples of this application are in the 'weathering' (exposure to sunlight) of wool fibres and the oxidation of synthetic polymer fibres, e.g. polyvinyl chloride (PVC).

If fibres are stained with selective fluorochromes, the sensitivity of the fluorescence microscope investigation can be greatly increased. Fluorochromes with affinities for particular chemical species may be applied by conventional staining procedures and, when examined under UV excitation, can show the location and proportion of the compound under study. Much has been learned, for example, about the position and structure of the ortho- and para-cortices in animal fibres by the use of acridine, rhodamine and geranin fluorescent stains.

During processing and finishing, various additives are applied to textile fibres, many of which fluoresce. Most spinning lubricants are oil based and will fluoresce under UV excitation, as will certain antistatic agents and easy-care finishes. Because most, if not all, domestic washing powders contain optical brightening agents (OBAs), it is possible, using a fluorescence microscope, to tell whether or not a fibre or fabric has been washed and whether the OBA was fibre specific.

With all fluorescence studies it is most important that control samples are also examined, to ensure that the fluorescence observed is coming from the specimen and is not due to an artefact, e.g. fluorescence of the mounting medium or secondary fluorescence from the autofluorescence of neighbouring fibres.

6.2 Infrared microscopy

In the same way that coloured materials selectively absorb white light, most organic and some inorganic molecules will absorb certain frequencies of IR radiation. The wavelengths which are transmitted or absorbed are determined by the chemical bonds present in the material, and the resulting IR spectrum is characteristic of the compound and may be used for identification. IR spectrometers operate by irradiating the specimen under study with broad-band (typically 2–15 µm) IR radiation and plotting the percentage of radiation absorbed against the frequency. The spectrum obtained, either by direct transmission or attenuated total reflection, is characteristic of the material and, by comparing it against reference spectra, the substance may be identified. Recently, the technique of Fourier-transform IR (FTIR) spectroscopy has been introduced and, when coupled to a microscope, allows the IR absorption spectrum from extremely small samples, e.g. single fibres, to be obtained.

The most obvious difference between a conventional light microscope and an FTIR microscope is in the optics. Because IR radiation is absorbed by glass, reflecting (or Cassegranian) objectives have to be used, in which the magnification (and focusing of the incident IR beam) is achieved by reflection rather than refraction. The modern commercial FTIR microscope consists of an IR source, beam splitters, an interferometer for wavelength control, the microscope, IR detector and spectrometer, and a computer for processing the signals generated. IR microscopy is used in fibre identification, particularly forensic studies, and is also applied in the analysis of textile dyestuffs, oils and other processing auxiliaries.

A complimentary technique to FTIR microscopy is that of microspot Raman spectroscopy, in which the minor proportion of IR light which undergoes a change in wavelength on interacting with a material is measured. While IR light is sensitive to the polar nature of molecules, the Raman shift is sensitive to the covalent bonds of materials. Thus, information not obtainable by FTIR microscopy may be gathered by Raman microscopy and vice versa.

Both FTIR and Raman microscopies are highly specialized techniques of analysis which may be applied to textiles. For further information on these methods, the reader is advised to consult a standard reference work on the subject, e.g. that by Banwell (1983) or Willard *et al.* (1981).

6.3 Confocal microscopy

Although at the time of writing no significant applications of confocal microscopy specific to textiles have been reported, the potential capabilities of the technique warrant inclusion in a book of this nature. In a conventional compound light microscope, the specimen is illuminated by a cone of light from the condenser. After interacting with the specimen, all the rays which are collected by the objective proceed to form the primary and secondary images: in effect, a magnified version of the area illuminated. This means that light from above and below the focal plane of the objective also contributes to image formation, and out-of-focus parts of the specimen appear in the final image. The principle of the confocal microscope is that instead of a whole field being illuminated and magnified as a complete area, the specimen is scanned by a finely focused beam or simultaneous array of beams. The illuminating aperture which forms the beam(s) is positioned in such a way that it is in common focus with an imaging aperture, i.e. the apertures are confocal. Because these apertures are in common focus, light coming from above or below the illuminating focal plane is effectively out of focus when it reaches the imaging aperture and thus does not contribute to the confocal image. This results in an image being formed of a single optical section of the specimen which is essentially free from out-of-focus blur.

There are two main types of confocal microscope: those which build up the image point by point by scanning a diffraction-limited spot of laser light across the specimen, so that the complete image is formed by the TV system of the instrument; and those which scan the specimen with many thousands of pinholes, usually in a spinning disc, which is confocal with and synchronized to an imaging set of pinholes. This latter confocal system is generally referred to as a tandem scanning reflected light microscope (TSRLM) and operates in real time, while the laser point scanning instrument is termed a confocal laser scanning microscope (CLSM) and accumulates the complete image electronically.

The characteristics of confocal images differ from those of conventional compound microscope images due to both the restriction of image-forming rays to those in focus and the sectional build-up capability offered by image storage. Some of the potentially beneficial applications of this type of microscopy to textiles are given below.

(i)   Non-invasive serial sectioning (opto-digital microtomy). Many cross-sections prepared with the plate and Hardy microtomes (Section 5.1), although adequate for identification and structural studies, are of insufficiently flat field to allow precision photomicrography. With a confocal microscope, however, a section of varying thickness can be thin sectioned optically and fibre cross-sections of extreme fineness can be viewed with greatly improved resolution, without mechanically changing a standard specimen.

(ii) High resolution epi-fluorescence. The use of a diffraction-limited laser spot as exciting radiation limits the size of the illuminated volume element (voxel) and controls the area of excitation to a far greater extent than is the case with conventional fluorescence microscopy. The reduction of secondary fluorescence 'glare' and restriction to a single plane of imaged fluorescence greatly improves resolution and clarity. A three-dimensional fluorescence image can be built up with successive scans in the CLSM.

(iii) High resolution sectional imaging. By making the confocal apertures sufficiently small and rejecting out-of-focus blur, both the lateral (X, Y) and depth (Z) resolution of a confocal microscope can be significantly increased over conventional systems.

(iv) Non-destructive examination of surface topography. Images of opaque specimens may be built up with successive confocal scans and may be colour or intensity coded by the image-processing computer. This overcomes the need for coating and examination under vacuum in the SEM, although of course the shorter wavelength of electrons means that the SEM will always have a greater theoretical resolving power than a light microscope.

There are many other potential applications for confocal microscopy in the study of textile fibres and those mentioned here are merely suggestions. It is hoped that the reader with access to a confocal system will be encouraged to experiment and apply this relatively new technology to the field of textiles. Many possibilities for useful development exist.

References

Banwell CN. (1983) *Fundamentals of Molecular Spectroscopy.* McGraw-Hill, New York.
Willard HH, Merritt Jr LL, Dean JA, Settle Jr FA. (1981) *Instrumental Methods of Analysis.* Wadsworth Publishing Co., Belmont, CA.

7 Scanning Electron Microscopy

The chief advantage of any type of electron microscopy lies in the greatly increased resolution which it can provide compared to that which can be achieved with light microscopy. The increased resolution is gained at the expense of greater cost and complexity of operation including the need to operate in a high vacuum. The SEM should really be compared with the stereo light microscope or the reflected light microscope when its great advantage of a good depth of field will be appreciated. This is complemented by quick and easy specimen preparation in the majority of cases which is to be compared with the difficulty of specimen preparation for the TEM.

The distinguishing feature of this type of electron microscopy is that the specimen is scanned raster fashion by a finely focused beam of electrons. The image is then formed from the radiation which is emitted as a result of this bombardment of electrons. Any of the resulting emissions may be used to form an image providing that a suitable detector is fitted to the microscope. However, the different types of emission do not all give the same resolution to the final image but can give different sorts of information about the specimen structure. In practice, secondary electrons are the usual choice for image formation as they give the maximum possible resolution, a detector for these being a standard fitment on an SEM. The resolution of the image formed is affected by the size of the focused electron beam and the area from which the radiation comes. It is important to note that, in all cases, the radiation is emitted from a thin surface layer so that no information is obtained about the underlying structures. This means that the SEM is an instrument for looking at surface appearance only.

One consequence of the electrons being absorbed into the surface of the specimen is that there is a build up of charge in non-conducting specimens. This can affect the behaviour of the secondary electrons so disrupting the image. As most fibrous specimens are non-conducting, they usually have to be given a thin coat of conducting material in order to obtain a steady image. Therefore, the image which is seen is actually that of the conductive coating.

7.1 Specimen–beam interaction

When the incident electron beam interacts with the atoms of the specimen under examination, a number of different types of emitted particles or radiation are produced according to the type of interaction. When electrons strike the nucleus of the atom a small number of them are reflected back through a large angle but the majority are deflected through smaller angles without losing much of their energy in the process. These electrons will then go on to strike further atoms as they travel deeper into the specimen, at the same time spreading out from the original point of contact. Some of these electrons will also find their way back to the surface by multiple scattering. These elastically scattered high-energy electrons, with energies up to that of the incident beam, are called backscattered electrons. Because of their high energy, they can emerge from deeper in the specimen and from a larger area than that struck by the beam, as shown in *Figure 7.1*.

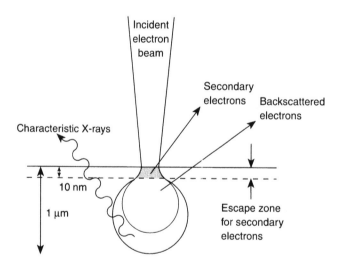

Figure 7.1:   The escape zones for the different types of radiation produced when an electron beam strikes a specimen.

If the high-energy electrons avoid the nucleus but collide with the electrons in the outer shell of the atom, some of these outer-shell electrons are ejected and an X-ray can be produced as a consequence. The ejected electrons have low energy compared to the incoming electrons and are called secondary electrons. Because of their low energy, the ones which reach the detector must have come from the top layer of the specimen, although they are produced throughout the interaction volume. The X-rays can escape to the surface from the whole of the interaction volume.

7.2 Secondary electrons

When the high-energy electrons in the incident electron beam strike the specimen they are gradually brought to a halt by repeated collisions with the atoms which make up the specimen. The volume in which this interaction takes place increases below the surface because the scattered electrons themselves go on to collide with other atoms. This repeated scattering gives rise to a pear-shaped interaction volume, as shown in *Figure 7.1*.

Only electrons with high energy are capable of escaping from the deeper parts of the specimen and being collected by the detector. Low-energy electrons which have been ejected from the atoms comprising the specimen are known as secondary electrons and are produced throughout the interaction volume. However, because of their low energy, they can only escape from the thin surface layer of the specimen. This reduction in penetrating power means that their area of origin is comparable to the area irradiated by the electron beam so providing the highest resolution images that the SEM is capable of.

Because they have low energy, these secondary electrons can be easily deflected by an electric field. This property is made use of in the standard SEM electron detector in which an electric field is used to deflect those scattered electrons into the collector which would not have entered otherwise. This gives rise to a high overall signal and also provides some signalling from those areas which are out of direct line of sight of the detector. Unfortunately, the secondary electrons can also be easily deflected when the specimen becomes charged, which can give rise to image distortion. When the beam is scanned close to corners or thin edges, extra secondary electrons can be produced because the interaction volume impinges on the second surface, as shown in *Figure 7.2*, so giving the appearance of extra brightness in these areas.

7.3 Backscattered electrons

Backscattered electrons are those which have been elastically scattered by the atomic nucleus, losing only a small amount of their energy in the process. Repeated scattering within the bulk of the specimen results in some of them emerging from the surface travelling in a direction such that they can enter the standard electron detector or can be collected by

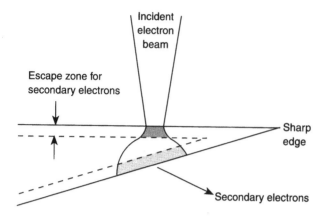

Figure 7.2:   Extra secondary electrons produced at a sharp edge.

a specially designed backscattered electron detector. They possess high energy, comparable in magnitude to that of the incident electrons, so that they are not easily deflected by either charge build up on the specimen or by the electric field used to deflect electrons into the standard electron detector. Their high energy also means that their point of origin can be deeper in the specimen where the interaction volume is larger. Thus, the possible resolution is not as high as that from the corresponding secondary electron signal which arises from the surface layer, as shown in *Figure 7.1.*

Backscattered electrons are detected by the normal electron collector only when they are travelling directly towards it so that in most circumstances the number collected is low and the signal correspondingly weak. The signal is also highly directional as areas which face away from the collector give no signal at all thus leading to an image with high contrast. Purpose-made backscattered electron detectors collect from a much larger solid angle than the standard electron collector in order to give a better signal-to-noise ratio.

The number of backscattered electrons emitted by a sample increases with the accelerating voltage used so that it is often advantageous to operate the microscope with a higher accelerating voltage than when using the secondary electron signal to form the image. The number of electrons scattered towards the detector also increases with the size of the atomic nucleus and hence with the atomic number of the sample. In an uncoated sample, this will give rise to contrast between areas which contain atoms of different atomic weight.

The backscattered electron image does not have the bright areas at corners or thin edges which are characteristic of the secondary electron image nor is it influenced by charging of the specimen; this can be seen by comparing the images in *Figures 7.3* and *7.4*. These factors mean that it is possible to obtain an image of the sample with a more even 'illumination' and without any areas which are too bright. The main disadvantage of the backscattered electron image is the lower resolution compared with the secondary electron image.

Figure 7.3:   Wool fibre: secondary electron image. Accelerating voltage 20 kV. Bar = 50 μm.

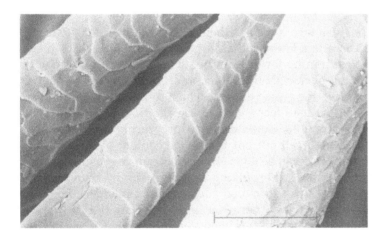

Figure 7.4:   Wool fibre: backscattered electron image (same sample as in *Figure 7.3*). Accelerating voltage 20 kV. Bar = 50 μm.

7.4 Resolution

Certain SEMs are capable of a resolution of 3 nm (1.5 nm with field emission source) but this is strongly dependent on using an ideal specimen. In general, the operation of the SEM requires a trade off between the various operating parameters so that the maximum resolution the microscope is capable of is not always present in the image.

The ability of the SEM to resolve fine structure is limited by the diameter of the electron probe used because it is not possible to resolve any detail which is smaller than the size of the probe. However, it is not sufficient to reduce the probe size in order to increase the resolution because the information which is gained from the sample is dependent on the number of electrons which are scattered by it which, in turn, is dependent on the number of electrons which are contained in the illuminating beam. Unfortunately, the number of electrons in the probe is directly related to its size because, in producing a fine probe, the optical system selects the centre of the probe and rejects any electrons outside the selected area.

The spot size, and hence resolution, is dependent on the accelerating voltage. This is because the field in the electron gun between the cathode and anode focuses the beam to a virtual source which is then imaged by the lens system to form the scanned spot. The size of this source changes with the accelerating voltage so that the geometry of the gun can only be optimal for a particular accelerating voltage, generally the highest one. At lower accelerating voltages the resolution is, therefore, below its potential maximum. Manufacturers often provide a taller anode for use with lower accelerating voltages in order to improve the resolution in this region.

When a smaller spot size is used without changing the intensity of the electron source there is a lower signal-to-noise ratio. This is a result of the smaller spot containing fewer electrons which in turn gives rise to fewer secondary or backscattered electrons being ejected from the specimen. In practice, this gives an image which contains a lot of random noise, making it difficult to focus it or to see the fine detail in it. The problem can be overcome to some extent by integrating the signal over time: the noise cancels itself out but the image content of the signal is additive. This integration can be done either electronically using a frame store or, as is more usual, when taking a photograph by scanning the spot more slowly across the specimen such that the photographic film acts as the integrating device. There is a limit, however, to this process because the requirement for the specimen to remain in one place during the process becomes very stringent. For high resolution work, it is necessary to start out with a brighter electron source, possibly provided by using a field emission source or a lanthanum hexaboride source.

7.5 Depth of field

Depth of field is one of the big advantages which the SEM possesses when compared with the light microscope, even for images at similar magnifications. It can be defined as the allowable variation in specimen height, on either side of true focus, within which the specimen appears to remain in sharp focus. The reason that the electron microscope has a greater depth of field than the light microscope is that the electron beam has a very low angular spread so that the spot size does not vary rapidly with height. The angular spread of the beam can be kept to a minimum by using smaller final apertures or by operating at greater working distances.

7.6 Charging

Most textile materials are non-conductors. This gives rise to the most serious problem for SEM with these materials, which is that of charge build up. If the electrons which strike the specimen are unable to flow to earth and thus complete the electrical circuit, the specimen accumulates surplus electrons in the area which is being scanned, so becoming charged. This charge will increase with time until it is high enough to discharge to earth. This process of charging and discharging will be repeated while the beam remains focused on that area. Pointed areas and edges concentrate the charge, giving rise to bright areas in these parts of the image, as can be seen in *Figure 7.5*. Discharge causes a sudden release of secondary electrons which results in a bright streak in the image. The presence of charge on a specimen may cause the incident beam to be deflected, giving the appearance of movement in the image. It can also cause adjacent charged areas to move apart by electrical repulsion if they are not firmly fixed to the stub.

Figure 7.5: Acrylic fibres showing charging effects due to insufficient coating. Bar = 200 µm.

The level of charge on a specimen is determined both by the energy of the electrons and the number of them striking the specimen in a given time. This in turn is determined by the accelerating voltage, spot size, gun brightness and speed of traverse of the incident electron beam.

The effects of charging can be overcome by coating the specimen with a conducting layer and ensuring that this is electrically continuous with the metallic mounting stub. The effects of charging on the image can also be reduced by using backscattered electrons to form the image as these are less affected by charge build up than are secondary electrons.

7.7 Beam damage

In the SEM, the energy contained in the electron beam is mostly dissipated as heat at the point of contact with the specimen. This can cause damage to the material being examined: for instance, the melting of thermoplastic polymers or the breaking of covalent bonds in long-chain molecules leading to depolymerization. The amount of heat generated in the specimen depends on the accelerating voltage, the beam current and the area which is scanned. As the magnification is increased, the area scanned by the beam becomes correspondingly smaller thus concentrating the heat input. The heat damage is also affected by both the thermal conductivity of the specimen, which is generally poor in non-metallic specimens, and the scanning time: the longer the beam dwells on one area the hotter that area will become. Metallic coating of the specimen helps to conduct away the heat besides limiting the depth of penetration of the beam.

7.8 X-ray microanalysis

When an electron is ejected from an atomic orbital by the incident electron beam, the atom is forced into an unstable high energy state. It can then relax back to a stable state by filling the position of the missing electron from another orbital and simultaneously emitting an X-ray photon with an energy equal to the difference between the energy levels of the two orbitals. The electron shells around an atomic nucleus have specific energy levels and are designated K, L, M etc., K being the one nearest to the nucleus with the lowest energy. If the electron which has been ejected came from the K shell, the X-ray radiation which is produced when it is replaced is termed K radiation with a subscript (α, β, γ etc.) which shows whether the replacement electron came from the next outer shell (L) or the one beyond it (M). Each shell has multiple electron orbitals each with their own energy level so that the possible transitions will have a range of different energies. Therefore, most elements will have multiple lines in each band of radiation but not all will be detected as the intensity of the lines varies. The various orbitals of the atom are all involved but the net result is that the X-ray spectrum produced contains a set of lines whose energies are characteristic of the atomic species involved. Elements can then be identified by measuring either the energy of the X-rays produced or their wavelength, which is inversely proportional to the energy. The heavier elements emit a larger number of X-rays because they have a greater number of orbital electrons. The wavelength of the emitted X-rays decreases with increasing atomic number of the target element, that is the energy increases. The energy of the electrons in the incident beam must be higher than the energy level of an X-ray if it is to be produced. In practice, considerably higher voltages are needed to give the most efficient excitation of spectral lines. SEM gun voltages of 20–30 kV are used to obtain the 1–10 keV range of X-rays where some emission lines from most elements can be found.

There are two types of X-ray detectors which can be mounted on electron microscopes, the wavelength-dispersive spectrometer and the energy-dispersive spectrometer. The wavelength-dispersive spectrometer uses Bragg diffraction by crystals to separate the X-rays in terms of their wavelength. Usually, a set of four selectable crystals is needed to cover all the elements. Rotation of the diffracting crystal allows each X-ray wavelength to be focused in turn on a suitable detector. Collection is inefficient because of the poor solid angle of collection and absorption losses in the analysing crystals. This means that beam currents have to be relatively high. It is more sensitive for light elements and the detection limits are better than for energy-dispersive analysis.

The energy-dispersive spectrometer uses a silicon detector which is cooled to the temperature of liquid nitrogen and which gives a current flow each time it is struck by an X-ray. The size of this current is proportional to the energy of the incident X-ray. One detector serves for all the elements, and the resultant signal is sorted out into an energy spectrum by a multichannel analyser, which means that a complete analysis can be obtained quite quickly. A typical spectrum is shown in *Figure 7.6*. The collection efficiency is high. A window usually made of beryllium has to be placed between the source and detector to prevent contamination and this places a limit on the lowest atomic number element which can be detected.

7.9 Measurement in the SEM

Despite its high resolution, the SEM is not an easy instrument to use for measuring objects. The problem lies in the fact that the magnification changes because the working distance alters as the microscope is focused.

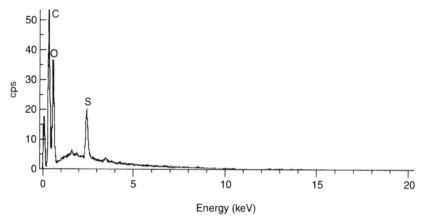

Figure 7.6: Energy-dispersive X-ray spectrum of wool.

This means that parts of the same specimen which differ in height will be imaged at different magnifications as they are individually brought into focus.

The built-in magnification display of the SEM should not be relied upon, however well it has been calibrated, as lens settings are subject to hysteresis and therefore do not necessarily return to the same values as they had when the calibration was carried out. To measure objects with any accuracy in the SEM, it is first necessary to use a suitably accurate calibration standard, as any measurement cannot attain a greater accuracy than that of the calibration used. Secondly, the specimen and standard should both be viewed at the same working distance and at the same magnification with no alteration of the focus in between. This can be achieved by first focusing on the specimen and then bringing the standard into view and focusing on it using only the height (Z) control. For really accurate work it is best to have the standard and specimen in view at the same time; this is usually achieved by depositing small polystyrene latex spheres of a known size on to the specimen.

7.10 Specimen preparation

Generally speaking, very little work is required to prepare textile specimens for examination in the SEM. This can be compared to the large amount of work which is needed to prepare a specimen for examination by the TEM. In many cases, all that is needed is to stick a small representative sample to a specimen stub. As nearly all textile specimens are good electrical insulators, the major requirement is to provide a conductive pathway to earth to disperse the incoming electrons.

7.10.1 Coating

There are two aspects to providing a pathway for the electrons to earth:

(i) bonding the sample to the metallic stub;
(ii) coating the whole assembly with a thin conductive layer.

Traditionally, the specimen has been bonded to the stub using a conductive adhesive such as colloidal silver or graphite. The drawback to this method is the time which is required for the solvent to evaporate from the adhesive. Residual solvent vapour will interfere with the achievement of a high vacuum both in the coating unit and the SEM. The use of a conductive adhesive does not necessarily improve the removal of electrons as it is the conductive coating which traps the electrons and consequently it is the continuity between this and the metallic stub which is important. In practice, the use of double-sided adhesive tape or adhesive tabs which can be coated instantly appears to work satisfactorily. However, it is possible that volatile components from these may evaporate off in the high vacuum environment of the microscope, giving rise to contamination in the column.

The main requirement for a coating material is that it should have a high electrical conductivity together with a grain size which is sufficiently fine such that it does not limit the resolution which is required from the specimen. The coating also affects the electron scattering properties of the specimen. Secondary electron emission is higher for metals than for non-metals and it also increases to some extent with the density of the metal. The number of backscattered primary electrons increases with the atomic number of the coating. A coating of higher atomic number also decreases the depth of penetration of the electron beam thus improving the spatial resolution. The most commonly used coating material is gold which has a high secondary emission coefficient and good granularity. Platinum/palladium and gold/palladium are used when a smaller grain size is required. There are three main ways of coating a sample with a conductive film, described as follows.

Sputter coating. In this method, the metal to be coated is bombarded with positive ions from an argon plasma which sputters the metal atoms in all directions. The plasma is produced by applying a high voltage (1–3 kV) between a metal anode and the sample holder which forms the cathode, the whole sample chamber being filled with argon at low pressure (0.1 torr). The argon atmosphere is produced by simultaneously pumping out the chamber with a rotary pump and bleeding argon into it. The presence of gas molecules means that the metal atoms cannot always follow a straight path to the specimen; some of them are deflected by the gas molecules so coating the specimen from different angles. This means that all facets of the sample can be coated in one operation, which is the great advantage of this method and the reason why it is the one most commonly used. The film deposition rate depends on the power input which is in turn dependent on the gas pressure and the applied (variable) voltage. The film thickness can be controlled by setting the coating current and timing the coating cycle.

The potential problems of this method of coating include heating of the specimen by the electrons which are generated in the plasma. This can cause problems with low-melting-point polymers such as polyethylene and other temperature-sensitive specimens. The problem can be minimized by coating using low deposition rates and by increasing the coating time accordingly. The coating is then built up from a number of thin layers with a pause for cooling between each layer. *Figure 7.7* shows a partially oriented polyester fibre which has been coated at a high current; this is to be compared with *Figure 7.8* which shows a similar sample which has been coated at a lower rate of deposition. Some sputter coaters

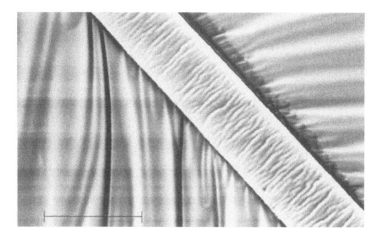

Figure 7.7: Partially oriented polyester fibre coated at too high a rate of deposition. Bar = 50 µm.

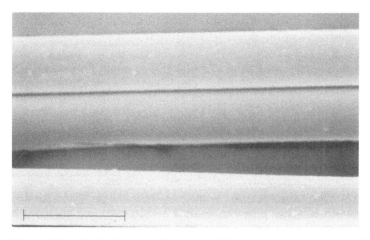

Figure 7.8: Partially oriented polyester fibres coated at a low rate of deposition. Bar = 50 µm.

are provided with a 'cool' attachment in which a magnetic field is used to deflect the heat-generating electrons away from the specimen stage.

Metal evaporation. In this method, the metal to be evaporated is heated in a high vacuum environment by passing a high current through a wire basket or boat containing the metal. The boat is made from a suitable high-melting-point material such as tungsten or molybdenum. The metal evaporates off in the high vacuum in the form of a spray which travels in all directions from the source. Carbon can also be evaporated in a similar manner.

The disadvantages of this method are the high cost of the equipment due to the two-stage pumping process needed to reach high vacuum, the long time taken to pump the chamber down to the required vacuum and the fact that the metal is only sprayed in a direct line from the source so that only those parts of the specimen which face the source are coated.

Carbon coating. This type of coating is usually used for specimens which are to be X-ray analysed but which need a conductive coating, as carbon does not interfere as much with X-ray analysis as a metal coating would do. The carbon can be deposited from either a carbon fibre braid or from two carbon rods, sharpened to a point, which are spring loaded against one another. In either case, a very high current is passed through the carbon which heats part of the carbon to white heat so evaporating some of it. This process can be carried out either in a conventional high-vacuum evaporation plant or in a coater similar to a sputter coater with only a rotary pump to provide the vacuum.

7.10.2 Cross-sections

These can be prepared in a similar manner to that of the plate method (Section 5.1.1) used in light microscopy. A specially prepared sample stub is required with one or more holes drilled through it (1-mm diameter is satisfactory). A bundle of fibres is pulled through the hole using a thin strong thread or wire. The fibres need to be packed as tightly as possible as this gives support against the cutting forces involved. The fibres are then cut flush with the stub surface using a fresh razor blade or scalpel blade, the quality of the cut governing the quality of the final image. The cross-sections so obtained must be given a conductive coating in a similar manner to all other fibre specimens.

The drawback to this method is that the knife blade distorts the ends of the fibres and the imperfections of the blade are all too evident with the increased resolution of the SEM, as can be seen in *Figure 7.9*. Unfortunately, embedding of the fibres as in sectioning for the TEM does not overcome the problem because the purely surface view of the SEM is not able to distinguish between the fibre and the embedding material.

Figure 7.9: Cross-section of trilobal nylon fibres. Bar = 100 µm.

7.10.3 Peeling

Peeling a fibre is one way of revealing some of its internal structure. The idea behind it is that the line of the peel will follow a natural line of weakness in the fibre, passing between the main structures. Some fibres, which are highly oriented with weak lateral bonding or have naturally microfibrillar structures, can be easily peeled whereas other fibres do not separate along the length.

To peel a fibre it is necessary to glue the lower half of it firmly to a specimen stub. The fibre is then viewed in a stereo light microscope so that it can be cut half way through at an angle with a small sharp scalpel blade. The part of the fibre thus cut free can then be grasped with a fine pair of tweezers and pulled slowly so as to peel it away from the lower half.

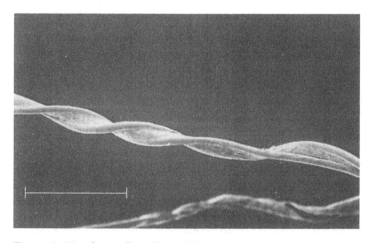

Figure 7.10: Cotton fibre. Bar = 100 µm.

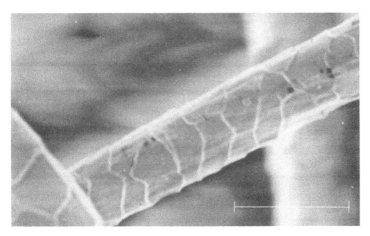

Figure 7.11: Wool fibre, untreated. Bar = 50 µm.

Figure 7.12: Wool fibre, after shrinkproof treatment. Bar = 50 µm.

7.11 Application of SEM to textiles

The applications of the SEM to textiles are mainly confined to the study of their surface topography, as illustrated in *Figure 7.10*, because the technique is not suitable for studying the internal structure of fibres. Surface features are of importance in those applications which involve either modification of the fibre surface or the application of a resin or other finish to the surface. Both of these features are found in the shrinkproof treatments which are used for wool, in which its scale structure is modified and a resin treatment is applied to it. *Figure 7.11* shows the untreated wool surface and *Figure 7.12* shows the surface after treatment with chlorine and a resin finish.

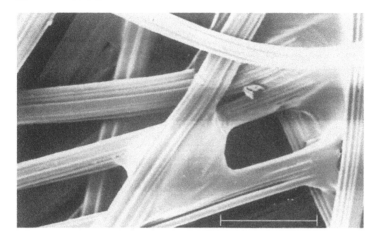

Figure 7.13:   Non-woven fabric showing fibre-to-fibre bonding of viscose fibres. Bar = 50 µm.

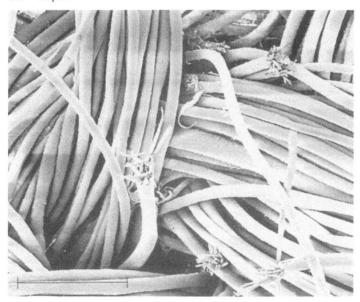

Figure 7.14:   Fabric showing fibre damage after abrasion. Bar = 200 µm.

Other uses of the SEM are:

(i)   the examination of the fibre-to-fibre bonding which is found in the manufacture of non-woven fabrics (*Figure 7.13*);

(ii)   the examination of fibre damage (*Figure 7.14*);

(iii)   the examination of the fine surface detail found particularly in natural fibres where the scale structures of animal fibres are used as the sole means of identification;

(iv)   the examination of fibre cross-sections, particularly in the case of synthetic fibres, where cross-sections other than circular are deliberately manufactured in order to change the properties of the fibres (*Figure 7.9*).

The SEM is not in general a very useful instrument for fibre identification as any internal features such as medullae and pigment are not visible when using it.

Further reading

Chapman SK. (1986) *Working with a Scanning Electron Microscope.* Lodgemark Press, Chislehurst, UK.

Goldstein JI, Newbury DE, Echlin P, Joy DC, Fiori C, Lifshin E. (1981) *Scanning Electron Microscopy and X-ray Microanalysis.* Plenum Press, New York.

8 Transmission Electron Microscopy

The main advantage of the TEM, which it shares with the SEM, is its greatly increased resolution when compared to that of the light microscope. Depending on the type of specimen being examined, the TEM is able to resolve structures comparable in size to inter atomic distances. The extra advantage that the TEM possesses over the SEM is its ability to examine the internal structure of textile fibres.

The TEM also shares with the SEM the disadvantages of greater cost and complexity of the equipment when compared to light microscopes and the need to operate with the specimen in a high vacuum. However, the biggest drawback of the TEM is its need for a very thin specimen, preferably less than 100 nm thick, so that the electron beam can pass through it. This requirement means that specimen preparation for the TEM is difficult and time consuming, ruling out the use of transmission electron microscopy for routine examination of fibre samples. Its main use is thus as a research instrument.

8.1 Electron–specimen interaction

One of the problems of the TEM is the lack of contrast in the image. In order to understand and overcome this problem, it is necessary to study the interaction of an electron beam with the specimen. There are three possible types of interaction of an incident electron with the specimen:

(i) the electron passes through the specimen without being deflected;
(ii) the electron is elastically scattered losing no energy in the process;
(iii) the electron is inelastically scattered with consequent loss of energy.

In a thin specimen made up of low-atomic-weight material, as most textile fibres are, the majority of the electrons pass through without any interaction. However, electrons passing straight through a specimen contain no information about that specimen as there has been no interaction with it. Electrons need to undergo a scattering interaction in order to impart any information about the material to the final image. Furthermore, if the image is to contain structural details of the specimen, the different areas of the specimen have to scatter the electrons by varying amounts in order to obtain contrast. Electrons which are scattered either collide with the walls of the microscope or are deliberately excluded from the imaging system by the use of a small aperture in the objective lens. This loss of electrons leads to contrast in the final image because areas which have scattered some of the electrons appear darker than areas where the electrons have passed straight through, due to the absence of the scattered electrons from that part of the image. The smaller the objective aperture that is used the greater is the contrast, but there is a trade off with a consequent reduction in resolution. Lack of contrast is a serious problem with fibre samples as they have essentially the same atomic weight throughout the material.

The amount of scattering which takes place increases with thickness, atomic weight and crystallinity of the specimen. Any overall increase in thickness of a specimen causes a general increase in scattering from the whole of the specimen so that the contrast between the various areas of the specimen remains unchanged. One way of increasing contrast is to stain samples with a heavy metal, such as mercury, uranium or osmium, which attaches itself preferentially to areas in the specimen having a particular chemical structure. This approach has been more useful with the natural fibres, which have a greater range of reactive chemical structures than do synthetic fibres. Synthetic fibres have few structural details which can give rise to contrast in the image, the main differences being in degree of crystallinity and orientation of the polymer chains.

Electrons which have lost energy through being inelastically scattered reduce the resolution of the image because they represent chromatic aberration. The chance of an electron losing energy increases with the thickness of the specimen.

8.1.1 Effect of accelerating voltage

TEMs, apart from special high-voltage instruments, generally operate with accelerating voltages of between 40 and 120 kV. Biologists use the lower accelerating voltages whilst material scientists favour higher voltages. The higher the accelerating voltage the greater is the energy of the electrons in the beam and also the shorter is their wavelength. The greater energy of the electrons leads to greater penetration but with consequent lower contrast as they are then deflected less by the specimen. The shorter wavelength means a higher theoretical resolution. If the highest resolution is not required and the specimens are sufficiently thin, it is possible to use a lower accelerating voltage such as 80 or 60 kV to improve the contrast.

8.2 Electron diffraction

Electron diffraction is very similar to X-ray diffraction in that, when a beam of electrons is focused on a crystalline material, parts of the scattered wave will reinforce each other strongly in specific directions thus forming a diffraction pattern. In the TEM, the diffraction pattern is formed in the back focal plane of the objective as part of the normal imaging process. The conditions for reinforcement are given by Bragg's law:

$$2d \sin \theta = n\lambda$$

where d is the distance between the diffracting planes, the Bragg angle θ is the angle between the direction of the incident beam and the diffracting planes, λ is the electron wavelength, and n is a whole number. In the TEM, the wavelength of the electrons is very small and dependent on the accelerating voltage ($\lambda = 0.0037$ nm for 100 kV accelerating voltage) so that the angles involved are also small and $\sin \theta$ can be replaced by θ to give

$$2d \theta = n\lambda.$$

When the sample is a single crystal the diffraction pattern will take the form of a set of spots with each one corresponding to a set of planes in the crystal. The presence or absence of particular spots depends on the orientation of the crystal. With a sample which is a powder or which consists of many small crystallites, the individual crystals may be randomly arranged or they may be oriented to a greater or lesser extent. The randomly arranged sample will give a diffraction pattern which consists of rings instead of spots but with all the expected diffraction spacings being represented. Partially oriented samples will give segments of rings and the shorter the arc of these the greater is the degree of orientation of the crystallites in the sample. In this type of sample, not all reflections will be present. In the case of fibres, the crystallites are aligned parallel to the fibre axis but are randomly arranged at right angles to this. Therefore, only certain diffraction arcs are present in the pattern in the fibre axis direction (meridional reflections) whereas all the expected diffraction arcs are present in the pattern perpendicular to the fibre axis (equatorial reflections).

In the TEM, a magnified version of the diffraction pattern can be viewed on the screen by altering the excitation of the intermediate lens. The area of the sample which gives rise to the diffraction pattern can be selected by means of an aperture which is inserted into the plane of the first intermediate image whilst the specimen is viewed in the normal mode (*Figure 8.1*). Because the magnification of the diffraction pattern can be altered, it is necessary, if the atomic spacings of an unknown sample are to be determined, to calibrate the microscope using a sample of known atomic structure.

Figure 8.1:   Diffraction pattern of Kevlar fibre.

The diffraction pattern, because it represents the regular atomic structure, is very sensitive to beam damage. If the diffraction pattern can be obtained from a fibre it will be seen to fade rapidly as it is viewed.

8.3 Dark field

Dark-field images are a way of increasing contrast by using only the scattered electrons to construct an image. They can be produced by intercepting the transmitted electrons and allowing only scattered electrons to reach the image plane. Dark-field imaging is possible with any specimen but it is more useful for crystalline materials which produce brighter images due to the coherent scattering of the beam. Dark-field images can be produced either by offsetting the objective aperture so as to intercept the undeviated part of the beam or by leaving the aperture central and tilting the beam so that the central spot is blocked by the aperture.

Displacement of the aperture means it is no longer symmetrically disposed around the optic axis of the microscope thus introducing unwanted aberrations into the image and, because of this, the beam tilting method is preferred for high-resolution dark-field microscopy. The manipulation of the beam tilt controls is carried out while the diffraction pattern is in view so that the appropriate diffraction spot can be manoeuvred so as to be visible through the objective aperture whilst the central spot is blocked by the aperture. A very small objective aperture may be used in order to ensure that the dark-field image is produced from specific diffraction spots. The use of a small objective aperture, however, can reduce the potential resolution and also cause problems due to rapid contamination build up on it.

In fibres the dark-field image is mainly dependent on the crystalline material present (see *Figure 8.2*); therefore, it is rapidly degraded by beam damage.

Figure 8.2: Dark-field image from meridional reflections of Kevlar fibre. Bar = 5 μm.

8.4 Beam damage

The electrons which make up the beam in a TEM are highly energetic: at 100 kV accelerating voltage each electron has an energy of 100 keV. These electrons possess sufficient energy to break the covalent bonds holding together the long-chain polymers that make up the fibres. Therefore, while the fibre is being examined in the TEM, its structure is being continually destroyed. Certain fibres, such as those containing a large number of aromatic groups, are more resistant to beam damage than polymers which are predominantly aliphatic such as polyethylene. This effect can be seen by viewing the diffraction pattern of a fibre in the microscope when it will be seen to quickly fade and disappear. The smaller structure in the fibre, particularly the crystalline structure, is destroyed before the larger structural entities.

The damage caused is proportional to the number of electrons which strike the specimen such that it depends on both the intensity of the beam and the length of time the beam remains focused on the specimen. The number of electrons per unit time striking the specimen is proportional to the square of the magnification assuming that the brightness is kept constant. This means that beam damage increases sharply with magnification.

Unfortunately, there is no way of avoiding beam damage as a certain number of electrons have to strike the specimen in order to extract the necessary information for the image at a given resolution. Reducing the intensity of the electron beam means that the exposure time of a photograph has to be increased. The determining factor of the quality in an electron micrograph is the 'noise' caused by the random nature of the electrons rather than the grain size of the photographic emulsion such that a fixed number of electrons are required to give a certain quality of photograph. Photographic emulsions are considered to be near 100% efficient for electrons; that is, each electron striking the emulsion results in one grain being developed. Therefore, a higher magnification image taken using a fast emulsion may have on it only the same information as a lower magnification image taken using a slower emulsion. The use of a fast emulsion, however, may be dictated by other factors such as the need to focus at a higher magnification.

In order to minimize the effect of beam damage it is first necessary to determine the total electron dose which destroys the features of interest. Once this is known then all the instrumental adjustments including focusing are carried out away from the area of interest so that the total electron dose can be used solely for photographic recording. The available electron dose sets an upper limit to the magnification and time of exposure that can be used.

Lattice fringe images are the most sensitive to beam damage and can only be obtained from fibres which are stable to the electron beam, such as carbon fibres or (briefly) aramid fibres. Dark-field images at low magnifications can be obtained from many fibres if the crystalline regions are sufficiently large. Diffraction patterns can be obtained at lower electron intensities such that most fibres should give one.

8.5 Specimen preparation

The most important requirement for a specimen which is to be examined with the TEM is that it is thin enough to allow the electron beam to penetrate it; this means preferably less than 100 nm thick. The finer the structure that is required to be examined the thinner the specimen needs to be. This means that it is impossible to examine whole fibres as they are too thick, so some means has to be found of preparing thinner pieces of the fibre. There are two ways of doing this: either the fibre is broken down into smaller pieces or it is cut into thin sections. Of these methods, sectioning is by far the most useful technique as any internal structural relationships in the fibre are retained during preparation.

8.5.1 Fibrillation

This process is only possible with certain fibres, either those composed of smaller fibrils, such as wool, or those which are highly oriented and whose lateral bonding is weak. The method depends on using sufficient physical force, such as ultrasonic waves, to break down the fibre into small pieces. These can then be deposited straight on to a prepared specimen grid and examined in the electron microscope. The method is thus fairly simple but suffers from the problem that only small fragments of the fibre are viewed without any indication of from where in the fibre they came.

8.5.2 Sectioning

Most of the relevant information about fibre structure lies in the relationship between the structural elements and their disposition within the fibre; for instance, the difference between the skin and the core of the fibre. In general, the fibre structure does not vary along its length but changes markedly across its diameter. For this reason, it is of more value to examine thin sections of the fibre, either longitudinal sections or cross-sections, although this method of preparation requires time and skill.

The fibre cannot be sectioned directly because it is impossible to hold it firmly enough to stop it being deflected by the cutting forces. It is therefore necessary to embed the fibre in a resin of suitable hardness and section it using an ultramicrotome capable of cutting sections down to 10 nm thickness with a glass or diamond knife.

8.5.3 Embedding

Before cutting thin sections with an ultramicrotome it is necessary to embed the fibres in a suitable resin. This has to be of the correct hardness to enable sections of between 20 and 100 nm to be cut, and also to be stable to the electron beam. A large number of formulations for embedding media are available but either standard or low-viscosity epoxy resin will be suitable for most fibre requirements. These resins usually consist of four components of which one is a plasticizer which allows the correct degree of hardness to be obtained. Epoxy resins need to be cured in an oven at around 60°C so it is important that heating does not affect the fibre structure. The fibres should be degreased before embedding in order to improve the adhesion of the resin.

Fibres need to be supported in some kind of mould while the resin sets. If only cross-sections of the fibre are required, small circular polythene capsules can be used. These are slit half-way through and the bundle of fibres wedged in the slit so that the fibres run down the central axis of the capsule. If the capsule is then placed in a tightly fitting hole, it will hold the cut edges together so that the embedding resin will not run out. However, it is easier to use flat aluminium foil trays to embed the fibres, cutting out the sample with a saw when the resin has set. This allows any orientation of the fibres to be chosen at will. If the fibres are held extended on a cardboard frame this will hold them clear of the bottom of the mould while the resin sets.

8.5.4 Ultramicrotome

For cutting thin sections with an ultramicrotome either glass or diamond knives can be used. Most fibres can be cut with a glass knife, although a diamond knife is necessary for cutting hard fibres such as Kevlar. The glass knives are freshly prepared from glass strips using a knife breaker. Only a very small area of the specimen can be cut at once (in the region of 1 mm^2) so that the specimen block has to be trimmed to a truncated point containing the area of interest. A small trough containing distilled water is attached to the triangular glass knives; diamond knives have the trough built in. The sections float off on the water surface from the knife edge in the form of a ribbon and they can be picked up by lowering the edge of the specimen grid into the water. The thickness of the sections can be judged by the interference colours which they show.

Sections are compressed by the cutting forces which are involved in producing them, as can be seen in *Figure 8.3*, and consequently there is the possibility that the orientation in the fibre structure is also affected by these forces. The sections can be stretched by exposing them to chloroform vapour or heat but this will not necessarily reinstate the fibre structure.

8.5.5 Support grids

Whatever form the specimen takes, whether it is a section or fragment, it must be supported on a thin film when it is examined in the TEM, as the samples are generally too small to be supported by even a fine mesh grid on its own. The electron beam has to pass through this film which detracts from the image quality. Therefore, the film must be as thin and structureless as possible. The film is itself supported by a 3-mm diameter copper mesh grid so that it only has to be strong enough to support itself between the bars of the grid. The finer the mesh of the grid the better is the support but large areas of the specimen can be obscured if there are too many grid bars; the most frequently used grids are the 200-mesh type. The film covering the grids can be either a plastic, such as formvar, or carbon.

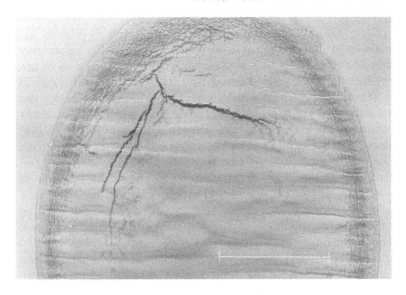

Figure 8.3: Oblique cross-section of aramid fibre impregnated with silver sulphide (TEM). Note striae and cracking caused by cutting forces. Bar = 10 μm.

Plastic films are made by dipping a clean microscope slide into a dilute solution of the plastic (e.g. 0.2% formvar in chloroform) and allowing it to drain in an upright position. The edge of the film is then cut round and the slide lowered at an angle into a petri dish containing distilled water when the film should float off on to the surface of the water. The grids are coated with the film in either of two ways. They can be placed directly on to the floating film and a filter paper placed on top. When this has soaked up water it is possible to lift it out with the grids and film adhering to it. These are then left to dry in a dust-free atmosphere. Alternatively, the grids can be placed on a piece of wire mesh at the bottom of the dish before the film is floated on to the surface. The water level is then reduced, usually by using a petri dish with a tap fitted, which causes the film to be lowered on to the grids.

Carbon films are stronger and more beam resistant than plastic films which allows a thinner film to be used. They are prepared by evaporating carbon on to a freshly cleaved mica surface. The evaporation is carried out in a vacuum coater which is capable of passing a high current through carbon rods at a suitably high vacuum. The current heats the contact area of the pointed carbon rods to a high temperature thus evaporating the carbon. The thickness of the carbon film can be monitored by having alongside the mica a piece of white glazed porcelain which has a drop of high vacuum oil on it. The thickness can then be judged by the colour of the porcelain, the oil drop remaining uncoated. The carbon film is floated from the mica and coated on to the grids in the same manner as described for the plastic film. Carbon can also be coated on to plastic-coated grids either to strengthen the film or as a route to carbon-coated grids if the plastic is subsequently dissolved.

8.5.6 Staining

Staining is a way of increasing the contrast of specimens by selectively incorporating heavy metals into the structure. Most of the stains in use are non-specific and have been derived from those in use in biological microscopy. Stains are typically solutions of salts of heavy metals such as osmium tetroxide, uranyl acetate and lead citrate. This type of stain reacts with some components of the structure leaving a deposit of heavy metal in those areas. Specific stains can sometimes be designed to react with known chemical compounds in the fibre so demonstrating at the microscopical level the location of the chosen compound. An example of this is shown in *Figure 8.4* where methyl mercury iodide has been used to stain the sulphur-containing amino acid cystine in wool. Such stains depend on there being available a staining compound which can react with the required groups.

Staining is easier to carry out on the cut sections than the whole fibre because of the time needed for the stain to penetrate through the whole fibre. Sections already mounted on grids can be either floated face down on the staining solution or completely immersed in it. After staining for the required time, they are washed in distilled water and dried on filter paper.

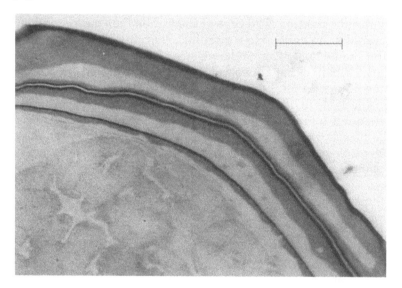

Figure 8.4:   Cross-section of merino wool stained with methyl mercury iodide showing the high sulphur-containing areas. Bar = 2 μm.

Further reading

Chescoe D, Goodhew PJ. (1984) *The Operation of the Transmission Electron Microscope.* RMS Handbook No.2, Oxford University Press, Oxford.

9 The Identification and Quantitative Analysis of Animal Fibre Blends

9.1 Introduction

Ever since the introduction of the Wool Products Labelling Act in the USA in 1939, the majority of industrialized countries have enforced legislation which makes declaration of the fibre content of all textile products mandatory. The justification for this requirement is considerable. Textile fibres vary enormously in price, performance characteristics and prestige value, and there are now so many different fibre types which may be blended together that the tactile properties of a textile product no longer offer a reliable indication of the fibres from which it has been made. Commercially, clear and accurate labelling of fibre content allows the consumer to make informed decisions on the potential purchase of textile goods, and to judge the relative value of articles made from different fibres or blends of fibres. The authenticity of a stated composition is therefore vital to both retailer and consumer.

For the reasons given above, the fibre compositions of all textile products, particularly of those made from blends of different fibres, are regularly checked. These checks may be made at any stage in manufacturing: from raw material, through card sliver, roving and yarn to fabric. Blend ratios often change with mechanical processing, as finer and less dense fibres are preferentially lost. The manufacturer needs to know what input levels result in the desired final composition, so that consistency can be maintained. At the retail level, the fibre content of all articles must be determined before the labels can be added and, even when fully labelled and stated, the accuracy of fibre content declarations are frequently checked by consumer protection organizations. A considerable number of technologists are employed by independent testing laboratories to carry out fibre content determinations according to national or international standards and the results, issued either as reports or certificates, are regarded as highly valuable documents. It is interesting that a practice which has been law for decades in the textile industry has only recently begun to be extended to other products, with manufacturers of foods, detergents, paints etc. now declaring their compositions for the benefit of the consumer.

9.2 The quantitative determination of fibre composition

The general procedure for quantitatively analysing a blend of fibres is based on the sequential chemical dissolution of individual components and calculation of the respective proportions of each by mass difference. Standard procedures apply for each fibre type (BS 4407 (1988), ISO 1833 (1977)) but the general principle is as outlined below.

1. Identify the fibre types present in the material.
2. Take two separate representative test specimens of minimum mass 1.0 g.
3. Determine the clean dry mass of each specimen to the nearest 0.0001 g.
4. Treat each specimen with the appropriate solvent to dissolve out the most soluble fibre, leaving the other fibres unaffected.
5. Filter the insoluble fibres into tared sintered glass crucibles, dry and weigh.
6. Treat each specimen with the solvent to dissolve out the next most soluble fibre.
7. Repeat steps 5 and 6 until only one fibre type remains.
8. Calculate the mass of each fibre type in the material by subtracting the weight of the crucibles and weighing bottles and the fibres dissolved at each stage.
9. Calculate the percentage of each fibre type in the material and adjust for moisture regain values. Quote both the individual duplicate and arithmetic means of the results.

Clearly, the above method is based on the different chemical solubilities of specific fibre types, and is basically a straightforward laboratory procedure capable of being carried out by any competent chemist. Where the blend to be analysed is composed of fibres which are chemically equivalent, however, this procedure obviously cannot be employed. It is therefore necessary to use other means to determine the blend's composition. This applies principally to blends of natural fibres, i.e.

mixtures of two or more vegetable fibres or two or more animal fibres.

In general, vegetable fibres have relatively distinctive and characteristic cross-sections, and it is possible to obtain a satisfactory analysis of blends of these fibres by using a combination of cross-sectional area examinations and measurements, together with longitudinal counting procedures, in a manner similar to that used for fibre diameter measurement (see Chapter 3). With animal fibres, the situation is more complex: apart from the fur fibres, which have characteristic ladder medullae, the cross-sectional and general appearance of animal 'wool' or 'hair' fibres is very similar. The principle features which allow the experienced analyst to distinguish between the different types of animal fibre are the characteristics of the cuticular scales, the range and uniformity of the fibre diameter, the crimp or lack of it, and the distribution and nature of any pigment present.

Mention has already been made of the use of scale casts in differentiating between wool and cashmere and the commercial importance of being able to make such a distinction (see Section 5.2). When the information required is quantitative, however, more complex and lengthy procedures must be employed.

9.3 The composition analysis of animal fibre blends

The basis of the accepted method of analysing blends of animal fibres is the identification of a statistically adequate number of fibres in a representative sample of the material in question. Numerical counting of the different fibre types, however, is not sufficient because the fibres will invariably be of different diameters. As the proportion by mass is required, the diameter and degree of medullation (if any) must also be determined for each fibre type. The pioneering work in this field was published by Wildman in 1954 and the technique has remained basically unaltered since then. Essentially, the method involves the identification and measurement of images of fibres using a projection microscope in the same way as that described for fibre diameter determination (Chapter 3). Statistical calculations and reference tables are consulted to establish the number of fibres which must be identified and measured and, when the coefficients of variation of the diameter distributions have been determined, the additional number of fibres which must be identified and counted only. These calculations and statistical procedures are complex and outside the scope of this book; however, the general formula for the determination of the proportion of a given fibre by mass in a two-component blend of animal fibres will illustrate the principle of the technique: % A, where A is the coarsest component of the blend and contains medullated fibres, and both fibre types are of equal specific gravity,

$$= \frac{[nA(\overline{d}^2A + \sigma^2A)] - [nmA(\overline{d}^2mA + \sigma^2mA] \times 100}{[nA(\overline{d}^2A + \sigma^2A)] - [nmA(\overline{d}^2mA + \sigma^2mA)] + [nB(\overline{d}^2B + \sigma^2B]}$$

where:
A = coarser component of the blend;
B = finer component of the blend;
n = number of fibres recorded;
$\overline{d}^2$ = the square of the mean fibre diameter;
σ^2 = the variance of the fibre diameter distribution;
m = medullated fibres;
$\overline{d}^2m$ = the square of the mean diameter of the medullae;
σ^2m = the variance of the distribution of medullae diameters.

While the projection light microscope is the original instrument intended for the application of the analysis, certain groups of workers, most notably at the Deutsches Wollforschungsinstitut (DWI) in Aachen, Germany, have adapted the technique to the SEM. The principle of the method, however, remains the same, i.e. identification of fibre type by appearance coupled with diameter measurement and counting.

9.3.1 Limitations

Although an apparently straightforward procedure, the accuracy of this microscopical technique is entirely dependent on the experience of the operator. The ability to recognize the characteristics which define a fibre as being of a particular type takes many years to achieve, and is a highly specialized skill. While a number of textile microscopists may feel confident enough to confirm the identity of a raw material sample of wool, cashmere or mohair etc., by comparing with reference samples, the quantitative analysis of an animal fibre blend in the dyed and finished state is an entirely different matter and should not be attempted by any person other than a recognized expert in the field. The number of microscopists who currently carry out quantitative analyses of animal fibre blends is very small indeed. It is also important that the fibres in the material to be analysed are in reasonable condition, as severe mechanical or chemical damage can erode the cuticles of animal fibres and prevent positive identification.

9.4 The current position

In recent years, the popularity of speciality animal fibres, particularly cashmere, has resulted in a greatly increased number of fabrics purporting to contain this fibre appearing on the market. While the higher quality products will in general declare their animal fibre compositions accurately, a considerable number of goods attempt to flout the fibre content labelling laws, and claim to either contain cashmere when they do not, or state that the cashmere content is higher than it actually is. As might be expected, there has been a corresponding increase in demand from manufacturers and consumer protection bodies for quantitative animal blend composition analyses. For the genuine specialists in the field, this situation simply presents an increased workload; however, there have been two other, very different, developments.

The first is that some laboratories which are not sufficiently experienced are carrying out analyses of animal fibre blends and issuing incorrect results. Their main error is that of distinguishing between cashmere and wool purely on the basis of fibre diameter, rather than identifying the fibres by their scale structures, uniformity of diameter, pigment distribution etc., as should be the case. Any fibre finer than a given value (usually 15 or 16 µm) is automatically classed as cashmere. The folly of this approach may be demonstrated by the fact that many fine lambswool fibres can be as fine or finer than certain cashmere fibres and there is often considerable overlap in the diameter distributions. In Australia, a special breed of merino sheep is kept indoors and produces ultra-fine wool. This wool, called Sharlea, has a mean diameter of approximately 15.5 µm and contains many fibres finer than cashmere. Its cost, however, is significantly less than that of true cashmere.

The second, cashmere-led development in this field has been the attempt to find a non-microscopical method of analysing animal fibre blends. The British Textile Technology Group have used their DNA analytical technique to establish the purity of samples of raw material, initially when the suspected contaminant has first been identified microscopically. This technique is under ongoing development and the intention is to use it to replace the microscopical method for the quantitative analysis of dyed and finished animal fibre blend fabrics. It is not yet known what will be the cost or accuracy of such tests.

Figure 9.1 shows examples of animal fibres viewed with the light microscope, while *Figure 9.2* shows the same samples viewed with the SEM. (Samples a, b, c in each figure were all taken from garments labelled '100% cashmere'.)

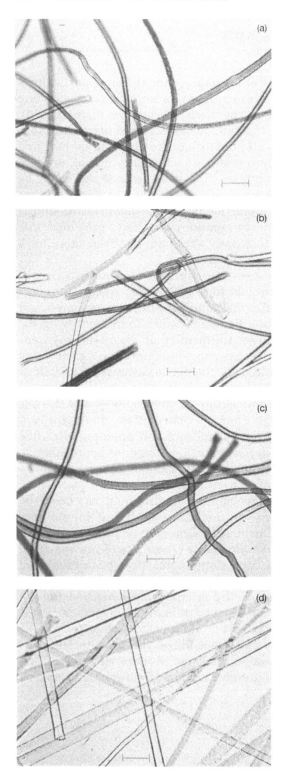

Figure 9.1: Light micrographs of animal fibres.
Bar = 100 μm.
(a) Pure cashmere;
(b) cashmere adulterated with wool;
(c) wool/nylon;
(d) mohair.

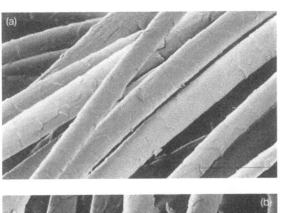

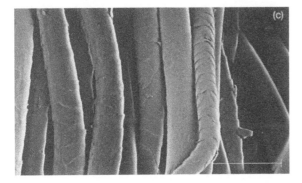

Figure 9.2: Scanning electron micrographs of animal fibres.
Bar = 50 μm.
(a) Pure cashmere;
(b) cashmere adulterated with wool;
(c) wool/nylon;
(d) mohair.

References

British Standard BS 4407. (1988) Methods for the quantitative analysis of fibre mixtures.
International Standard ISO 1833. (1977 (1980)) Quantitative analysis of fibre mixtures.
Wildman AB. (1954) *The Microscopy of Animal Textile Fibres*. Wool Industries Research Association, Leeds.

Appendix A
Glossary

Analyser: a polar/polarizer placed after the object (usually between the objective and the primary image plane) to determine optical effects produced by the object on the light with which it is illuminated.

Anisotropic: the possession of optical properties in a material that vary with the changing direction of propagation of light through it.

Birefringence: the difference between the two principal refractive indices (RIs) of a fibre.

Blend: a mixture of different fibres.

Bobbin: a cylindrical or slightly tapered former with or without a flange for holding slubbings, rovings or yarns.

Bulk: the space-filling properties of a textile product.

Card sliver: a continuous length of fibres produced from loose fibre by the carding machine.

Cheese: a cylindrical package of yarn wound onto a flangeless support.

Coefficient of variation (CV): a dimensionless property which measures the spread of a population distribution. $CV\% = \sigma \times 100/x$ where σ is the standard deviation and x is the arithmetic mean.

Composition: (fibre) the percentage by mass of different fibre types present in a textile.

Cone: a conical package of yarn wound on a conical support.

Continuous filament: a yarn composed of one or more filaments that run essentially the whole length of the yarn.

Core: the central part of a fibre.

Crimp: the waviness of a fibre.

Diameter distribution: the spread of individual fibre diameter measurements about the arithmetic mean of the measurements. Usually expressed as the coefficient of variation (CV%).

Fibre content: the percentage of each fibre in a blend.

Finishing: processing applied to a fabric after weaving.

Handle: the quality of fabric or yarn assessed by the reaction obtained from the sense of touch.

Hank: an unsupported coil of yarn.

Isotropic: the possession of optical properties by a material that do not vary with the changing direction of propagation of light through it.

Medulla: the central portion of some animal fibres consisting of a series of cavities formed by the medullary cells which collapse during the growth process.

Micron: a unit of measurement equal to 1000th part of a millimetre (abbreviation for micrometre).

Optical path difference (OPD): the distance of the displacement in air of two light waves emerging at right angles to one another from a birefringent fibre.

Orientation: a predominant direction of linear molecules in the fine structure of fibres.

Package: an assembly of yarn.

Polarizer: a device for producing plane-polarized light.

Polaroid: a specific type of polarizer in the form of thin plastic sheet (trade name).

Representative sample: a sample taken such that every individual in the parent population has the same chance of being included in the sample.

Roving: a continuous length of staple fibres of linear density suitable for spinning.

Sheath: the outer part of a fibre.

Staple fibre: fibres of finite length (as opposed to a Continuous filament).

Standard deviation: a calculated value which indicates the variability of a set of figures based on the deviations of individuals from the mean.

Textured: a continuous-filament yarn that has been processed to introduce durable crimps, coils, loops or other fine distortions along the length of the filaments.

Tow: an essentially twist-free assemblage of a large number of substantially parallel filaments.

Warp: threads lengthways in a fabric as woven.

Weft: threads widthways in a fabric as woven.

Yarn: a continuous length of either twisted staple fibres or untwisted continuous filaments, which may be woven or knitted.

Appendix B
Suppliers

Standards: British Standards Institution, 389 Chiswick High Road, London W4 4AL, UK. Tel (0)181 996 9000; fax (0)181 996 7400.

Projection microscope (IWTO Standard): R&B Instruments, Unit 3A, Farnley Low Mills, Bangor Terrace, Leeds LS12 5PS, UK. Tel. (0)113 279 1066; fax (0)113 231 9655.

Shirlastains (Textile Fibre Identification Stains): Shirley Developments International Ltd, PO Box 162, Shawcross Street, Stockport SK1 3JW, UK. Tel. (0)161 480 8485; fax (0)161 480 8580.

Index

Milton Keynes UK
Ingram Content Group UK Ltd.
UKHW031152141024
449569UK00024B/853